After the Finish Line

The Race to End Horse Slaughter in America

A Division of BowTie, Inc.
Irvine, California

To every person in America
who is doing anything to save horses
and end their slaughter.

Karla Austin, *Business Operations Manager*
Mark Simon, *Book Editor*
Jarelle S. Stein, *Associate Editor*
Nick Clemente, *Special Consultant*
Jill Dupont, *Production*
Laura Tucker, *Photo Research and Proofreading*
Susan K. White, *Book Design*
Maricel Quianzon, *Cover Design*

Photographs courtesy of: Rebecca Baucus, 95; Pierre Bellocq, 81; John Bellucci, 104; Carrie Devorah, 60; Dogwood Stable, 57; Peter and Mary Gregory, 164; Equine Advocates, 89, 112, 121; Bill Finley/Thoroughbred Times, 36; Fort Meyer, 139; Suzie Oldham, 52, 107; Flashpoint Photography, 134, 135; Stock Studios Photography, 41; Thoroughbred Times, 15; Robert "Skip" Trimble, 69; Lydia A. Williams, 129

BowTie Press®
A Division of BowTie, Inc.
3 Burroughs
Irvine, California 92618

Library of Congress Cataloging-in-Publication Data

Heller, Bill.
After the finish line: The Race to End Horse Slaughter in America / by Bill Heller.
p. cm.
ISBN 1-931993-60-2
1. Animal welfare. 2. Horses. I. Title.

HV4749.H45 2005
636.1'0832—dc22

2004017007

Printed and bound in the United States
10 9 8 7 6 5 4 3 2 1

Cover: Ferdinand winning the 1986 Kentucky Derby with Bill Shoemaker up. Photo by Tony Leonard.

TABLE OF CONTENTS

ACKNOWLEDGMENTS

This book literally would not have happened without John Hettinger, who showed an entire industry what one man can do to help make things right. His spirit and love of horses are pure and profound.

My editor and friend at *Thoroughbred Times,* Mark Simon, and Don Clippinger of *Thoroughbred Times* turned Hettinger's idea of a book into a reality, and I cannot thank them enough.

Chris Heyde of the Society for Animal Protective Legislation was incredible. Besides answering countless questions and sending me crucial material, he helped me understand the issue with more clarity.

I could not have written the chapter about Texas without the help of Skip Trimble of the Texas Humane Legislative Network. Carol Chapman, Mary Nash, and Mayor Paula Bacon of Kaufman, Texas, helped as well.

Susan Wagner of Equine Advocates, Cathleen Doyle of HoofPAC, Jo Anne Normile of CANTER, Michael Blowen of Old Friends, Gina and Curtis Brown of Spring Hill, and Pam Berg of the Glen Ellen Vocational Academy provided valuable background, and they collectively made this book easier to write and deeper in quality.

Thanks to trainer Nick Zito and his wife, Kim; Grace Belcuore, Diana Pikulski of the Thoroughbred Retirement Foundation; Russell Williams, Alan Leavitt, Michele Oren of the Thoroughbred Retirement Foundation's Exceller Farms; my dear friend Howard Nolan; Jim Trimble of the Thoroughbred Retirement Foundation; Priscilla Clark of Tranquility Farm; Richard Duchossois of Arlington Park; Barry Schwartz of the New York Racing Association; Jeanne Mirabito, Shon Wiley, Lori and Jack Neagle, and Kelly Lynn of ReRun; Jeffrey Tucker; Pierre "Peb" Bellocq; Herb and Ellen Moelis of Thoroughbred Charities of America; Judy Bokman of the Standardbred Retirement Foundation; Paula and John Campbell, Fred Noe, Annette Fisher of Happy Trails Farm Animal Sanctuary; Dot Morgan of New Vocations; Ward Stutz and Tom Persechino of the American Quarter Horse Association; Jay Hickey of the American Horse Council; Bob Curran of the Jockey Club; David Carr of the United States Trotting Association; Erin and Steve Porter; Peter and Mary Gregory, and Ellen Harvey with Harness Racing Communications.

U.S. Congressman Ed Whitfield interrupted his busy schedule to answer questions on several occasions, as did his wife, Connie, vice chairman of the Kentucky Horse Racing Authority, Tarrant County Assistant District Attorney Ann Diamond of Texas, and New York State Assemblywoman Deborah Glick.

Thanks to the copy editors at *Thoroughbred Times* who helped check facts, including Laura Tucker and Chris Bauer.

Thanks to my wife, Anna, and our son, Benjamin.

CHAPTER 1

Final Justice

The message flashed on Carol Chapman's computer screen on Sunday at 5:40 p.m., October 12, 2003:

> **HORSE NEEDS HELP!**
>
> *This alert is from a very good group in Pennsylvania. They spend a great deal of time at the infamous New Holland Livestock Auction rescuing horses. If anyone can help, please give Kelly a call. Her husband, Tracy, went to Texas to testify against the Betty Brown pro-slaughter bill.*
>
> *Thanks, Chris*

Chapman was used to such cries for help. The fifty-four-year-old financial reporting compliance specialist and author has her own horse rescue organization, The Last Refuge, in Hitchcock, Texas, and her own Yahoo message rescue forum on the Internet whose motto is: "Until the last horse is safe, we will not rest."

Kelly Young, who runs Lost & Found Horse Rescue Foundation out of York, Pennsylvania, and Chris Heyde, of the National Horse Protection Coalition based in Washington, D.C., were hoping to save one of the hundreds of horses who would be auctioned off the next day at the weekly sale in New Holland, Pennsylvania, where "killer buyers" make a living by purchasing discarded horses and trucking them halfway across the country to be slaughtered at one of two foreign-owned plants in Texas, which sell the meat for human consumption overseas.

The horse they were hoping to spare was an unlikely candidate for slaughter, an impeccably bred seven-year-old Thoroughbred named Monetary Justice, a grandson of the immortal Secretariat who had three productive seasons of racing in New York, winning four of thirty-two starts and earning $133,570. In the ensuing two years, however, the son of Academy Award did not even finish as high as third in fourteen starts and earned only $3,130.

He had become a liability, even after he left New York for easier competition in Maryland and Pennsylvania. In his final start, August 16, 2003, at Philadelphia Park Racetrack, Monetary Justice finished fifth by eighteen and a half lengths in a $4,000 claiming race—meaning any interested owner or trainer could have purchased Monetary Justice or any other horse in that race for that price. Nobody wanted him. Nobody except killer buyers.

It is unclear when Monetary Justice broke the sesamoid bone in his right front ankle, or how it was broken, or who put a bandage on that ankle when he showed up at New Holland that Sunday. The injury would not be a deterrent to the killer buyers who wanted to acquire him and then ship him to Texas to be slaughtered.

Chapman called Young to learn more about the horse. "Monetary Justice didn't belong there," Chapman said. "He stood out there head and shoulders. He exuded quality, even in the kill pen. There's something special about this horse."

Chapman had to act fast and she did. First, she wired $350 to a friend in Pennsylvania to cover the cost of buying Monetary Justice before the sale and paying for a mandatory Coggins test for equine infectious anemia, a viral disease without a vaccine or a cure, that is required before a horse can be shipped out of state.

On Monday morning, she wired more money to buy two other horses, a dark bay gelding and a dark bay mare, both of unspecified breed, for sale at New Holland. They would be joined by a fourth horse, a Thoroughbred named Quick Moment, who had been bought back by his original owner off the backstretch at Penn National Race Course in central Pennsylvania so he could be retired instead of slaughtered. Chapman's friend gathered the four horses in his trailer and drove them to his farm so they could be transported to Last Refuge in Texas.

Because she had done this many times before, Chapman had access to a transport network that provided the names of commercial transporters. "I knew there was one trailer up there getting ready to come back to Texas," Chapman said. "The driver and his wife were in Pennsylvania visiting her parents. They were looking for a run to come back. They live in Beaumont on the Louisiana border, about three hours from me."

Saving horses is costly. Chapman had to pay for temporarily boarding her horses, a fee for having them pulled out of the sale, the purchase price, and the transportation costs. "I paid three thousand dollars to move those horses down here," she said. "It was an expensive trip. It's personal money out of pocket. Nobody can save the world. We just pick our spots."

The trip to Texas the horses would have had to make if the killer buyers had purchased them was in stark contrast to the trip to Texas they received thanks to Chapman's intervention. Instead of being stuffed in undersized trailers meant to transport cattle, the horses were placed in individual stalls—Monetary Justice got a double-sized stall to prevent further injury to his ankle. Instead of nonstop travel without water, food, or rest, Chapman's horses were treated humanely. "I was in constant contact with the driver and his wife," Chapman said. "They got to Beaumont, and because it was very late at night, they unloaded the horses and let them rest for the night."

The horses arrived at Last Refuge the following afternoon with fourteen people eagerly waiting for them. "Most of the people there were Quarter Horse people," Chapman said. "When the first Thoroughbred [Quick Moment] walked off the van, they oohed and aahed and took pictures."

Quick Moment was taken to another rescue, Habitat for Horses, and placed in the farm's intensive care barn so his knee injury could be evaluated. He recovered and is enjoying his new life.

The gelding and a huge, seventeen-hand-tall mare were next off the trailer. Chapman had already picked out new names for them, Skip and Mary, to honor Skip Trimble and Mary Nash for their contributions to the battle to end horse slaughter in Texas. "Skip and Mary are two heroes down here," Chapman said.

Monetary Justice was last off the van. He was so undernourished that his ribs were visible. Chapman guessed he was four hundred pounds underweight. Yet he held his head high as he surveyed his new home. "He looked around like he was measuring everyone," Chapman said. "It was like, 'Okay, I'm here. I've arrived. You can bow down now.' "

Chapman reached out her left hand to greet him. Monetary Justice placed his nose to it. "He has the kindest eyes," Chapman said. "His eyes were soft and they were wide open. And we just looked at each other."

Monetary Justice won a race at Aqueduct in New York in 2001, earning a purse of $7,500 for his owner, but wound up in a slaughter auction two year later.

Chapman knew how close Monetary Justice had come to being a statistic, one of more than fifty thousand horses slaughtered at the two Texas plants in 2003.

"There is an essence of purity to horses," Chapman said. "Our relationship with horses is different than with any other animal. In many ways, dogs are like our children. They're companions. Our relationship with horses is not childlike. There is always a sense of self. A horse knows. He knows the difference between himself and other members of the herd and the difference between him and humans. For them to put that aside to work in partnership with you in riding or in racing is almost a gift that they're giving us.

"Because we entered into this relationship, we have a responsibility. They have made a commitment to us. We have to make a commitment to them. You can't discard that. It's not like, 'You're my car; you're three years old; I've got to trade you in.' We, as human beings, have lost that sense of continuity. We tend to discard things when we get bored with them. The commitment to a relationship with a horse is something we have to get back to. People can't turn their backs on them. I can't turn my back on them."

So she saves as many horses as she can. She picks her spots. And she talks to as many people as she can about stopping the slaughter of horses, while trying not to think about the horses that are not rescued.

"I can talk about this in public all day," Carol Chapman said. "Then I come home and throw up."

CHAPTER

2

Lost Overseas

When Dermott Meere, a horseman from Ireland who was working in upstate New York, decided to breed his mare Dawn of Life, he chose the stallion Academy Award, a son of Secretariat who was then standing at historic Claiborne Farm in Paris, Kentucky. Academy Award's female pedigree was impressive. His dam, Mine Only, was a daughter of Mr. Prospector, one of the most dominant sires in the history of Thoroughbred racing. The mating of Academy Award and Dawn of Life produced Monetary Justice. Academy Award's adjoining stall at Claiborne Farm belonged to 1986 Kentucky Derby winner and 1987 Horse of the Year Ferdinand. Monetary Justice would be spared from slaughter. Ferdinand would not. Sadly, Ferdinand was not the first Thoroughbred superstar slaughtered for meat overseas.

Exceller, who defeated Triple Crown winners Seattle Slew and Affirmed in the 1978 Jockey Club Gold Cup at Belmont Park racetrack, was slaughtered in Sweden in 1997. That should have been a wake-up call to the Thoroughbred industry. But it did not trigger a solution, for five years later, Ferdinand's life ended in a slaughterhouse in Japan.

There was, of course, an outpouring of sympathy and anger, but the ugly reality is that there were too many people who thought Exceller's and Ferdinand's lives meant nothing. Should their deaths mean nothing, too?

What happened to Exceller and Ferdinand is the same thing that happens to countless other Thoroughbreds, Standardbreds, Quarter Horses, Arabians, and Paints every day while horse slaughter continues at the three plants operating in the United States, two in Texas and one in Illinois.

Learning of Exceller's fate stunned his former North American trainer, the late Racing Hall of Famer Charlie Whittingham. "It's a sad thing ... you'd expect more from people," Whittingham told Mike Mullaney of the *Daily Racing Form,* Thoroughbred racing's past-performance daily newspaper. Seeing Exceller's name listed on a 1997 ballot for election to the Racing Hall of Fame—where he was ultimately inducted—prompted the editors at the *Daily Racing Form* to do a "Where are they now?" story. When the answer was "dead in a slaughterhouse in Sweden," Mullaney began researching. His enterprising stories in the *Daily Racing Form* in 1997 brought to light the sad fate of one of the world's great Thoroughbreds and jump-started the battle to end horse slaughter in America.

As a racehorse, Exceller was fabulous, not only in America but also in Europe; not only on turf, but also on dirt, compiling a record of fifteen victories, five seconds, and six thirds from thirty-three starts and earnings of $1,674,587. He was bred in Kentucky by Mrs. Charles W. Engelhard and owned by Belair Stud and Nelson Bunker Hunt.

Trained in France by Maurice Zilber through his four-year-old season before going into Whittingham's care, Exceller, a son of Vaguely Noble out of Too Bald, by Bald Eagle (ironically, Whittingham's nickname was the Bald Eagle), won four Group 1 turf stakes in Europe (Group 1 designates the best stakes races in Europe; Grade 1 designates the best stakes in North America)—the Prix Royal-Oak, Grand Prix de Paris, and Grand Prix de Saint-Cloud in France and the Coronation Cup in England. In the United States, he won two Grade 1 stakes on dirt, the Hollywood Gold Cup Handicap and the Jockey Club Gold Cup, plus four Grade 1 stakes on turf, the Hollywood Invitational Handicap, San Juan Capistrano Invitational Handicap, Sunset Handicap, and Oak Tree Invitational, in addition to the Canadian International Championship Stakes in Canada. Exceller was tough, versatile, and top class.

The defining moment of Exceller's career came in the one-and-a-half-mile Jockey Club Gold Cup on a sloppy track at Belmont Park on October 14, 1978. Then five, Exceller took on two Triple Crown winners, the four-year-old Seattle Slew and the three-year-old Affirmed.

In his two starts before the Jockey Club Gold Cup, Seattle Slew had defeated Affirmed by three lengths in the Grade 1 Marlboro Cup Handicap and beaten Exceller by four lengths in the Grade 1 Woodward Stakes, improving his career record to thirteen wins in fifteen career

Racing Hall of Fame member Exceller could not escape an ignominious demise in a slaughterhouse in Sweden.

starts. He would go off the strong 3-to-5 favorite in the Jockey Club Gold Cup while ridden by Racing Hall of Fame jockey Angel Cordero Jr.

Affirmed, who had fifteen wins and four seconds in nineteen career starts going into the Jockey Club Gold Cup, went off at odds of 2.20-to-1 under the great young rider Steve Cauthen while coupled in the wagering with stablemate Life's Hope. Exceller, ridden by legendary Bill Shoemaker, went off at 3.80-to-1 in the field of six.

Affirmed never had a chance as his saddle slipped. Cauthen could not control him, and so he, Seattle Slew, and Life's Hope battled head-to-head in a suicidal duel on the lead for the first three-quarters of a mile, which was run in a very swift 1:09.40. They still had another three-quarters of a mile to run, and Exceller was more than twenty-two lengths back in fifth place. Seattle Slew put away Affirmed and Life's Hope before Exceller flew up into contention on the far turn as if he would blow by Seattle Slew. Yet Seattle Slew, after losing the lead by a half-length, would not let Exceller get away and gamely fought back, losing at the wire by a nose in an unforgettable display of courage. Exceller remains the only horse to ever defeat two Triple Crown winners in a single race.

Exceller was sent to stud in 1980 at Gainesway Farm in Lexington, Kentucky, and sired several stakes winners, none of them as accomplished as himself. As demand for his services fell and he was no longer deemed to be commercially viable as a stallion in North America, Exceller was sold to Gote Ostlund, a Swedish owner and trainer of Standardbreds (harness horses), and exported to Sweden in 1991.

Mullaney's story revealed what happened to Exceller after he left the United States. Ostlund, who spoke limited English, told Mullaney that the twenty-four-year-old Exceller had died earlier in 1997 because "he was very old." But Mullaney got a more detailed account from Ann Pagmar, who owned a small farm in central Sweden where Exceller spent his final year after standing one season in Denmark. Pagmar said:

> Shortly after Exceller came to me last year, the owner called and told me to kill the horse because he couldn't pay for him. He said that since we weren't breeding Exceller, there was nothing else to do with him. The owner didn't want to pay for a stallion license and he wasn't paying me, so I offered to take Exceller, to buy the stallion license, and to breed him.
>
> He wouldn't give him to me, saying the horse shouldn't keep moving from one owner to another. The stud fees would help me get paid, but the owner was very stubborn. He wanted to kill the horse. I had him a year, and I became very close to him, like I do with all my horses. Over the next few months he [Ostlund] called several times, asking if I had killed Exceller yet, and I kept telling him I needed written authorization. He finally sent it to me in April.
>
> He told me to bring the horse to the slaughterhouse, and that's what I had to do. I walked him over myself. I made an appointment because I wanted to get it over with quick, but they were very busy when we got there and we had to wait. Exceller knew what was going on; he didn't want to be there. Standing with him like that … it made me feel like Judas.

Exceller was slaughtered on April 7, 1997. He was gone, but not forgotten.

A few years earlier, Barbara Kraus had already taken action to stop such practices. After reading an article in the weekly newsmagazine *Thoroughbred Times* about the deplorable conditions at auctions in New Holland, Pennsylvania, where killer buyers gather weekly to purchase horses for slaughter, Kraus became interested in finding an alternative.

She sponsored two horses that were rescued by the United Pegasus Foundation, a 501(c)(3) charity that finds second careers for California Thoroughbreds and other horses. To qualify as a 501(c)(3) charity, an organization must spend the money it raises on a tax-exempt activity, such as prevention of cruelty to animals.

Interested in doing even more, Kraus posted a message on an Internet forum, asking other people to join her in sponsoring horses, right about the time Mullaney's July 20, 1997, story in the *Daily Racing Form* brought Exceller's gruesome death to light. Twelve people immediately signed up to join Kraus, and they decided to create their own group, calling it the Exceller Fund. Membership has since grown to almost one hundred, with each member paying a minimum of $15 monthly.

Through the spring of 2004, thirty-two horses had been saved by their efforts. Eleven of them have been adopted; the others remain at the United Pegasus Foundation's California farm, where the Exceller Fund covers their expenses.

One of the adopted horses was a 1997 foal that had been sold to a slaughterhouse as a yearling. He was rescued in December 1998 and named Exceller's Legacy because he is a maternal grandson of the slain horse. The Exceller Fund and the Kentucky-based Old Friends also rescued Narrow Escape, a daughter of Exceller.

Another rescued horse was Thunder Regent, a career earner of more than $500,000 and a multiple stakes winner who competed in the 1992 Breeders' Cup Mile, finishing thirteenth in the field of fourteen behind winner Lure. Despite injuries, he was still racing at the age of thirteen in cheap races in Pennsylvania and New Jersey when the Exceller Fund bought him. Though Thoroughbreds are the Exceller Fund's first priority, the fund has rescued two Standardbreds, a Belgian draft horse, a Hanoverian, and a Paint.

People are making a difference. But not enough of a difference. It happened again with Ferdinand.

The news of Ferdinand's death in a Japanese slaughterhouse—which happened in 2002 but was not discovered in the United States until Barbara Bayer, a freelance writer for *The Blood-Horse,* a weekly Thoroughbred industry trade journal, broke the story in July 2003—drove the point home with a sledgehammer. How could the 1987 Horse of the Year, a horse trained by Whittingham and ridden by Bill Shoemaker, who won the 1986 Kentucky Derby and the 1987 Breeders' Cup Classic by a nose over Alysheba in one of the greatest stretch battles in Breeders' Cup history, wind up slaughtered in Japan?

"It's scary," commented Michele Oren, the manager of Exceller Farm in upstate New York, a haven for old or broken-down horses, to William Nack, an eight-time Eclipse Award winning writer, in an August 5, 2003, story for ESPN.com. "Kentucky Derby winners are not meant to be part of the food chain. No horse is. When I heard about Ferdinand's death, I thought, 'No, not again!' "

Ferdinand, bred by California oilman Howard Keck and raced in Keck's wife's name, won eight of twenty-nine starts and earned $3,777,978, retiring as the fifth-leading earner in Thoroughbred racing history. He went to stud in 1989 at Claiborne Farm, the place of his birth. Initially, his stud fee was $30,000 for a live foal, but he had little success as a stallion and was sold to Japan's JS Company in the fall of 1994. Ferdinand was a stallion for Arrow Stud on the northern island of Hokkaido from 1995 through 2000.

Attempts by the farm staff to place Ferdinand in a riding club failed, and he was sold to a horse dealer named Yoshikazu Watanabe and moved off the farm on February 3, 2001. Watanabe initially told Bayer that Ferdinand had been gelded and given to a riding club, but Ferdinand covered six mares in 2001 and two in 2002. When Bayer told Watanabe she wanted to see Ferdinand, Watanabe said, "Actually, he isn't around anymore. He was disposed of late last year."

Bayer learned that Ferdinand's registration in Japan was annulled on September 1, 2002.

Howard Keck Jr., whose father bred Ferdinand, told Judy Holland of Hearst Newspapers in a May 1, 2004, story that he tried to buy the horse from the Japanese to bring him home and retire him to a life of leisure only to learn that Ferdinand had already been slain for meat. "I was shocked, because in this country our culture was different, because we place great value on these retired treasures," Keck said.

Obviously, not enough.

CHAPTER

3

Horses in America

The impact of horses on world civilization is unquestioned. "A comparison of civilizations that had horses with civilizations that did not, soon makes it clear that human history, at least in Eurasia, would have been profoundly different were it not for the horse," Melinda Maidens wrote in "The Horse in Human History," an Internet work. "Most human activities can be categorized as war, travel, work, or leisure. The horse has been indispensable in all four."

This is especially true in America, where horses are woven into its history, culture, language, and everyday lives. Horses are so significant in American history that the passing of a president is marked by a funeral that includes a riderless horse with boots reversed in the stirrups to signify a lost hero. Words such as "Thoroughbred" and "horsepower" are commonly used, while politicians frequently compare their campaigns to horse races. Horses have been an integral part of the fabric of the growth of the nation, in taming the West, travel, communication, and agriculture.

Melinda Maidens wrote that horses are believed to have originated in North America, but the only ones that survived prehistoric times were the ones that crossed a land bridge near present-day Alaska into Asia. They were brought back to North America, first by explorers and then by immigrants.

Horses were so indispensable to the Spanish conquistadors that they brought them across the Atlantic Ocean to explore the Americas. Slings were constructed to allow the horses to sway with the roll of the ship and reduce the weight on their feet, but as many as half the horses often died on the voyage. The section of the Atlantic Ocean known as the "Horse Latitudes" was named for the innumerable dead horses thrown overboard into the ocean.

But enough horses survived to produce a score of generations that carved a unique niche in American history and affected the American way of life.

In early colonial America, horses were the only expedient means of travel and method to move goods a considerable distance. The lack of wide roads discouraged the use of carriages.

Mail service by post began on January 22, 1673, when Governor Francis Lovelace of New York dispatched a rider to Boston. Others followed, and a former Indian trail between New York and Boston became known as the Upper Boston Post Road.

The first horse-breeding center in America was in Rhode Island, which produced the Narragansett Pacer. Laws prohibiting racing on city streets date back to 1674 when the citizens of Plymouth, Massachusetts, enacted an ordinance to ban it.

The stagecoach, a way of travel in America for some two hundred years, was introduced by Jonathan Wardwell of Boston, whose coach first trekked from Boston to Providence, Rhode Island, on May 13, 1718.

Horses gave America's Revolutionary War forces a decided edge on the British, who had great difficulty in shipping horses to the colonies and resorted to stealing them from colonists. In battle, horses allowed revolutionaries to use hit-and-run tactics on the British. Men such as Henry "Light Horse Harry" Lee and Francis "the Swamp Fox" Marion became legends by using those tactics. The most legendary was Paul Revere, whose ride to warn American patriot leaders that the British were coming to arrest them on the night of April 18, 1775, is one of the first facts of American history children learn.

George Washington was typically painted on a white charger, but his favorite horse was a chestnut hunter named Nelson, who was a gift from the governor of Virginia, Thomas Nelson Jr. Washington rode Nelson on fox hunts for ten years before the Revolutionary War began and then into many campaigns during the war. Washington also owned and raced a Narragansett Pacer.

The first racecourse in the colonies was built on Long Island in 1668, and Virginia soon emerged as a center for racing and breeding.

Racing blossomed in Kentucky in the second half of the eighteenth century, and a racecourse was laid out in Lexington in 1789. Arguments among Kentuckians were frequently settled by a horse race.

Horses were crucial to America's growth in the early eighteen hundreds in all aspects of transportation of people and merchandise. When New York Governor DeWitt Clinton opened the Erie Canal in 1825, barges were pulled by horses. Railroads did not replace horses as much as they reassigned them to transporting goods and freight back and forth from the trains.

Throughout the eighteen hundreds, horse-related industries and crafts provided livelihoods for many, including harness and carriage makers, veterinarians, blacksmiths, farriers, grooms, coachmen, farmers, and ranchers. On April 3, 1860, the Pony Express was started to deliver mail. When coal was first mined in the United States, hearty Shetland ponies from northern Scotland were imported in great numbers to work in the mines in Kentucky, Ohio, Pennsylvania, and West Virginia. In some mines, ponies would be bred, born, and put to work without ever seeing the light of day.

Other horses became heroes, from Comanche, who survived Custer's Last Stand on June 25, 1876, to racehorses such as Dan Patch and Man o' War. After the Battle of Little Big Horn, a horse was found in a thicket with seven arrows in his body. That horse, Comanche, was a gelding owned by Captain Keogh, one of the officers under the command of General George Armstrong Custer of the Seventh Cavalry. There were no other survivors, human or equine, of the 264 troopers under Custer's command. Comanche's wounds were treated and he was loaded on a riverboat and taken to Fort Lincoln in the Dakota Territory, where he was nursed back to health. Once he recovered, Comanche was given free rein of the fort's grounds. The Seventh Cavalry's commanding officer insisted that Comanche was saddled for all official occasions, but that he could never be ridden again. Comanche became a national celebrity and his obituary appeared in newspapers around the country. Comanche was taxidermied after his death and exhibited at the museum of the University of Kansas.

Dan Patch was America's first racing superstar. Compiling a record of fifty-four victories from fifty-six starts through July 1902, the pacer was so popular that he toured the country in his private railway car to stage public exhibitions, when he raced not against other horses but against time. In 1903 at Memphis, he paced a mile in 1:56 1/4 hitched to a sulky, then, on the same day, paced a mile in 1:57 1/4 hooked to a wagon. Dan Patch's race times were far ahead of his time. It was not until the 1970s and 1980s that the best pacers and trotters were able to routinely race a mile in less than two minutes.

Dan Patch's celebrity was such that there was Dan Patch brand smoking and chewing tobacco, a Dan Patch washing machine, and even a dance, the Dan Patch Two Step. At home, Dan Patch lived in such a gigantic barn that it was dubbed the "Taj Mahal." Born in 1897, he died on July 11, 1916. The next day, his owner, M. A. Savage, died. Some ninety years later, the United States Harness Writers Association honors his memory by presenting the Dan Patch Award to all of the sport's champions.

Man o' War was born the year after Dan Patch died. The Thoroughbred legend's only defeat in twenty-one races came as a two-year-old in the 1919 Sanford Stakes at Saratoga to a horse named Upset—a result that solidified Saratoga Race Course's legend as the "Graveyard of Champions" and also added a new definition to the American dictionary, the word "upset" thereafter also meaning to defeat unexpectedly. Just as Dan Patch did, Man o' War ran so fast, including a mile in 1:35 4/5, one and one-sixteenth miles in 1:44 4/5, and one and three-eighths miles in 2:14 1/5, that his records stood for decades. He completed his undefeated three-year-old season and his brilliant career by beating the winner of the first Triple Crown, Sir Barton, by seven lengths in a match race.

Man o' War sired 1937 Triple Crown winner War Admiral, seven other champions, and sixty-two stakes winners. Thousands visited Man o' War at his owner Sam Riddle's farm before Man o' War died of a heart attack in 1947 at age thirty. His funeral was broadcast over radio and covered by the media from all over the world. Today, Man o' War's statue greets visitors to the Kentucky Horse Park in Lexington.

When movies and television began to fill people's idle hours, horses were among the icons. Cowboy stars were linked to the horse they rode: Gene Autry and Champion, Hopalong Cassidy and Topper, Roy Rogers and Trigger, and the Lone Ranger and Silver.

Some horses famous without a rider: Fury, Flicka, and Mr. Ed.

In 1973, Secretariat graced the cover of *Time, Newsweek,* and *Sports Illustrated* in the same week, an unprecedented triple crown of media coverage, before he won the Belmont Stakes to become the first Triple Crown winner in twenty-five years. Thirty-one years later, more than one hundred twenty thousand fans poured into Belmont Park and millions watched on TV to see if Smarty Jones could become the twelfth Triple Crown winner in American racing. He did not.

In 2003, Americans flocked to movie theaters to see *Seabiscuit,* the movie adaptation of Laura Hillenbrand's brilliant book that was nominated for an Oscar as best picture.

Other equine performers appeared in circuses, horse shows, and fairs. Millions and millions of other horses, however, were not performers. They spent their lives toiling anonymously in fields of grain or on wagon trails.

They, of course, never negotiated the contract. Who would speak for them?

After witnessing the abuse of harness horses, Henry Bergh founded the American Society for the Prevention of Cruelty to Animals (ASPCA) in 1866. He also helped pass the first law to protect animals in New York state. The ASPCA provided ambulances to rescue injured horses and maintained shelters for them. There are now more than seven hundred ASPCA groups in the country protecting animals.

In 1888, as part of the legacy of Anne Waln Ryerss, Ryerss Farm for Aged Equines opened just northwest of Philadelphia as a hospital for ill, aged, and injured animals. Ryerss accepted its first residents on June 11, 1889, and has served ever since as a haven for old hunters, ponies, and workhorses from the Philadelphia area. The farm expanded to take a dozen retired horses that had worked pulling fire engines on the streets of Philadelphia, then moved to its current home in Coventryville, Pennsylvania, in 1987. Its 383 acres today are home to more than one hundred aged, retired, or abused horses. About 180 acres of the farm are used to raise crops, including timothy, alfalfa, soybeans, wheat, and corn. Any crops not used for the horses are sold to generate income for the farm.

Eighty-three years after Ryerss took in its first horse, the United States Congress passed the Wild Free-roaming Horses and Burros Act in 1971 to protect them from slaughter. In December 2004, the estimated 32,000 wild horses in America lost that protection when a rider in the consolidated appropriations bill permitted excess horses and burros to be sold at auction "without limitation."

Thus the number of horses slaughtered in the United States by foreign-owned companies to produce meat for human consumption overseas is likely to rise. There were 50,564 horses slaughtered in the U.S. in 2003, and that number rose to more than 56,000 in 2004 after the reopening of the nation's third horse slaughterhouse in Illinois in June of 2004.

U.S. Representative Ed Whitfield of Kentucky, who has helped sign up more than 220 cosponsors in the House of Representatives for the American Horse Slaughter Prevention Act, is committed to seeing that bill become law. "Horses have never been part of the food chain in the United States," Whitfield said. "It's unconscionable that we're allowing this to happen."

CHAPTER 4

Reality

This is what happens in the United States every day that horse slaughterhouses remain open in Texas and Illinois: An average of 250 horses go through a weekly sale, widely known as the "New Holland Killer Sale," every Monday in New Holland, Pennsylvania. Between 25 and 40 percent of the horses are purchased by killer buyers, who send the horses on a nightmarish trip, frequently without food, water, or rest, to brutal slaughter.

At the sale, horses are packed together in small pens, with some of the horses in the auction house not having access to hay or water. The horses that are tied up outside have neither hay nor water, and in the summer months, have to stand in the hot sun for hours. A lot of fighting frequently occurs among the animals because they are tied up too close to each other or not tied at all.

"Sadly, when I visited New Holland in July 2004, it was immediately apparent that the owners, handlers, and the veterinarian there have little regard for federal and state laws providing for the safety and welfare of horses," said Connie Whitfield, vice chairman of the Kentucky Horse Racing Authority and wife of U.S. Representative Ed Whitfield, cosponsor of the anti-slaughter bill in Congress. "It would make you cry."

At New Holland, Connie Whitfield purchased a Percheron draft horse for $375 that she renamed Mr. Ed. "The vet [at the sale] said he was nine years old. He was twenty-one."

She also bought a mini-pony for $125.

When she took the horses back to Kentucky, her veterinarian told her the draft horse was shockingly underweight. "The vet at the sale turned a blind eye," Whitfield said. "My vet in Kentucky said he was as underweight as a horse could be.

"The most startling thing was the surreptitious sale of a painfully lame horse, which my friend spotted in the killer buyer pen. He had a disabling infection in his leg and could only stand on three legs. It took hours, but I convinced one of the sales managers, who was Amish, that the horse should be euthanized. My friends and I stayed there to make sure the horse was."

For the unlucky ones purchased by the killer buyers, the horses begin their ride to death across the country.

"Truckloads of horses purchased at the sales are sometimes hauled in double-decker rigs designed for cattle and hogs," Karen Kirsch wrote in the March 29, 2002, issue of *The Chronicle of the Horse.* "With a clearance of about five feet six inches, these rigs aren't tall enough to accommodate a horse. Horse trailers have a clearance of at least six feet six inches, so some equines in double-deckers are forced into unnatural and life-threatening postures."

The gross treatment of horses traveling to be slaughtered prompted Congress to pass the Commercial Transportation of Horses to Slaughter Act of 1996 to protect the horses from abusive conditions. The bill, however, did not go into effect until February 5, 2002, because it took more than five years to write the regulations.

Now, shippers must provide horses with water, food, and rest for six hours prior to being loaded for transport. Once loaded, horses cannot be shipped for longer than twenty-eight hours without being unloaded for six hours and given a chance to rest, eat, and drink. While in transport, the horses must be checked at least every six hours to ensure that no horse has fallen or is in physical distress. Trucks shipping horses must segregate stallions and aggressive horses from others, and they must provide enough room and be equipped with doors and ramps that allow safe loading and unloading. Horses that may not be transported include those under the age of six months, horses that are severely injured or sick, or those blind in both eyes.

Before loading, the owner-shipper must sign a United States Department of Agriculture certificate stating each horse is fit to travel. The certificate, which must accompany the horse to the slaughterhouse, includes the name and address of the owner-shipper, a description of the horse, the time, date, and place the horse was loaded, and a statement that the horse was provided access to food, water, and rest prior to being loaded. Offenders face a fine of $5,000, with each violation treated separately. Of course, that matters only if they are caught.

"Transportation to slaughter is still not humane, even with the new regulations," said Diana Pikulski, executive director of the Thoroughbred Retirement Foundation, which has been saving Thoroughbreds since 1982. "They watered the regulations down. They're not enforceable. I was a criminal defense attorney for fifteen years. They are never going to prosecute anybody. Anybody who thinks twenty-eight hours on a van is humane is fooling themselves."

Unsatisfied that the Commercial Transportation of Horses to Slaughter Act did not ban double-decker transport of horses, Vermont, Massachusetts, Pennsylvania, and New York passed their own laws to outlaw them. Plans to enact a similar bill in Indiana in 2004 was defeated by opposition from rodeo interests. The timing could not have been worse. Nearly thirty horses of fifty being shipped in double-decker cow trailers died when the trailer crashed in Dearborn County, Indiana, on September 15, 2004.

Pennsylvania's law is the strongest. Since taking effect in August 2001, enforcement has been relentless and violators have been prosecuted to the fullest extent of the law.

Even if the ride to the slaughterhouse is more humane, what happens when the horses reach their destination? According to a controversial white paper prepared by the American Horse Council, United States Department of Agriculture veterinarians are on hand to "meet and inspect each shipment of horses upon arrival at the facility."

Yet horror stories are abundant. Chris Heyde, executive director of the National Horse Protection Coalition, wrote of his personal experience in the Winter 2002 edition of the *Animal Welfare Institute Quarterly*:

> Located at the rear of the nondescript facility was a double-decker trailer fully loaded with horses. They filled both rows and were unable to stand normally, forced to keep their heads low ... When some were moved off the trailer, workers poked them with long fiberglass rods through holes on the side of the trailer. The horses, typically very sensitive animals, slid and fell down the ramp only to be whipped by another worker's rod. All of the horses at the facility exhibited fear typical of 'flight' behavior in horses, pacing in prance-like movements with their ears pinned back against their heads and eyes wide open.
>
> Once inside the building, more callous workers, standing high on the railing that lined the stalls, beat the horses on the nose, forehead, neck, back, or hindquarters to get them to move. This continued until they entered the kill chute.

> Two egregious acts of cruelty took place right in front of me. Running across the floor of the barn was a grate-covered drain about three feet deep. A section of the grate was missing in one of the stalls through which horses were being forced. Because they were crammed into a space and panicking, each horse fell into the open hole, unable to get out since the floor was wet and slippery.
>
> Workers continued to beat the horses until they were able to throw their bodies out of this hole. Due to the overcrowding and panic, a large male got his leg hooked over one of the upper rails. Again, workers proceeded to beat him continually until the horse lunged forward gouging his leg open on the solid metal fence, which forced his free leg off the rail. Federal law requires the presence of a U.S. Department of Agriculture inspector during slaughter, but an inspector was nowhere to be found.

Kirsch wrote about the final minutes before slaughter in her article in *The Chronicle of the Horse.* "Animals are 'stunned' by firing a captive bolt [similar to a pneumatic nailer] into the brain, not intending to cause death, but merely to cause insensitivity to stimuli," she wrote. "Losing consciousness takes about a minute. Then they're hoisted aloft by a chain around one back leg, bled, beheaded, eviscerated, and skinned."

From 1980 through mid-2004, approximately four million horses have been slaughtered for human consumption overseas, according to the Equine Protection Network.

It is important to distinguish between slaughtering live horses for human consumption and rendering dead horses for animal feed. Horses are not slaughtered alive for pet food in the United States, and have not been for some twenty years.

Rendering plants process dead horses. By law, those plants cannot haul any living animal. A horse's body can have presences of medications or disease and still be rendered because of the extremely high temperature in the vats at the plant. Rendered products include feed, fertilizer, tires, camera film, and animal feed.

Killing horses for meat is another matter, a deeply disturbing one. Slaughtered horses must be free of disease, drugs, and chemicals. Yet horses continue to be slaughtered daily at plants in Texas and Illinois.

Living in denial is easier than facing the reality of horse slaughter in the United States. Thoroughbred owner, breeder, and industry leader John Hettinger knows. He has been there.

"We all knew what went on, and we pushed it into our subconscious and hoped that our horses don't wind up that way," Hettinger said. "When I thought of it, I tried to think of something else. Then Exceller hit me right between the eyes."

Other people's eyes were opened on October 30, 2004, when the Breeders' Cup, a championship day of racing for Thoroughbreds that attracts horses from around the world, was held at Lone Star Park near Dallas, Texas, not far from two of the three horse slaughterhouses still operating in the United States. Both Beltex Corporation in Fort Worth and Dallas Crown Inc. in Kaufman, Texas, are owned by foreign companies that slay horses for human consumption overseas despite Texas laws that prohibit the selling, exhibition for sale, possession for sale, or transfer for sale of horsemeat for human consumption. The political battle to close those two plants continued through 2004.

On July 31, 2004, a full-page, anti-slaughter ad in the *Daily Racing Form* contained an open letter from Breeders' Cup founder John R. Gaines who wrote, "As you know, horses have never been a part of the food chain in our country."

Others share that point of view. "In this country we do not eat horses," said Skip Trimble, a Thoroughbred owner, lawyer, and vice president of the Texas Humane Legislative Network. He is at a loss to understand how the two horse slaughterhouses remain open in Texas, defying Texas state law.

"I don't know of any law anywhere where you permit a foreign country to come into our state, use our natural resources, our horses, and produce a product for foreigners," Trimble said. "If we do that, how are we going to stop people from killing dogs and cats and selling them to Korea and China? They eat dog meat and cat meat. I'm not taking issue with anybody's culture. I'm taking issue with our culture. We don't eat horses."

Another legislative battle was waged in Illinois in 2004 as the Belgian-owned company Cavel International, whose horse slaughterhouse burned to the ground on March 31, 2002, reopened two years later amid great controversy.

The foreign companies that operate the three horse slaughterhouses in America have found support from the most unlikely sources: the American Horse Council (AHC) and the two principal veterinary associations entrusted with protecting the welfare of the country's near seven million horses, the American Association of Equine Practitioners and the American Veterinary Medical Association.

The American Horse Council issued a February 2002 white paper entitled "Why the Horse Industry Opposes a Federal Ban on the

Slaughter of Horses for Food," calling horse slaughter a necessary evil and concluding that "a federally imposed ban is not in the best interest of the horse's welfare." The American Quarter Horse Association immediately endorsed that opinion, even as the American Horse Council quickly switched its official position to neutral following an immediate, intense, and hostile reaction from the horse community. Two years later, the American Horse Council's official position is still neutral, according to its president, Jay Hickey.

In the spring of 2004, while Representatives John Sweeney of New York, Ed Whitfield of Kentucky, and more than 220 cosponsors backed the American Horse Slaughter Prevention Act before the United States House of Representatives, and Senator John Ensign of Nevada, a former practicing veterinarian, sponsored a companion bill in the United States Senate, the two veterinary organizations opposed the legislation.

In a position paper sent to members of Congress, the American Association of Equine Practitioners (AAEP) stated, "[T]he slaughter of unwanted horses at processing facilities is currently a necessary aspect of the equine industry in order to provide a humane alternative to allowing a horse to continue a life of discomfort or pain and possibly inadequate care or abandonment."

Others disagree with every fiber of their being. "I don't need a D.V.M. attached to my name to know when a horse is scared or when a horse is in pain," Hettinger said.

Diana Pikulski, the executive director of the New Jersey-based Thoroughbred Retirement Foundation, said, "The AAEP is just being lazy. That's just looking for some cheap way out and it's not responsible. Second of all, horse slaughter is not humane by any stretch of the imagination."

The AAEP position defies simple logic. According to the United States Department of Agriculture, in 1989, 342,877 horses were slaughtered in the United States. In 2003, that number had fallen to 50,564, a decline of almost 300,000.

According to the United States International Trade Commission, the number of horses exported to Canada and Mexico has remained constant the past fifteen years. Yet, there have been no reports of massive neglect and abandonment of horses since the number of slaughtered horses decreased dramatically.

Incredibly, in that same position paper, the AAEP tried to sugarcoat the process of brutal slaughter, saying, "The AAEP believes that euthanasia at a processing facility is a humane alternative to a life of suffering, inadequate care, and possibly abandonment."

Horses are butchered at slaughterhouses, not processing facilities. And the doomed horses are not euthanized. Euthanizing them would either taint the meat or make it less valuable on the market. Rather, the horses are hit with a captive bolt gun that is also used on cattle. But horses have longer necks and different skull shapes than cattle and dodge the gun. Sometimes small pieces of the horse's skull get driven into the brain. Sometimes those horses, still conscious, fall down and get up again. "It doesn't happen every time, but it happens," said Chris Heyde of the Society for Animal Protection Legislation and of the National Horse Protection Coalition.

Thankfully, not everyone is in denial. Instead of accepting the ugly reality of horse slaughter in America, people like John Hettinger are battling to end it. He has spent an increasing majority of his time in the past five years doing everything he can to increase awareness of horse slaughter, develop viable alternatives to slaughter, and pass federal legislation that would outlaw in the United States the slaughter of horses for human consumption. Not just racehorses, all horses.

"With the skins off, they all look the same," Hettinger said. "There are two things that will stop me from seeing this through, bankruptcy or my demise."

Hettinger, who started Blue Horse Charities and donated Exceller Farm in upstate New York to the Thoroughbred Retirement Foundation, is not alone. There are heroes everywhere. They stretch from Florida, where seventy-five-year-old Peter Gregory and his wife Mary run the 240-acre, 122-horse Retirement Home for Horses, to the state of Washington, where Steve and Erin Porter and two 4-H clubs raised more than $1,200 to come up with the winning bid of $1,375 to buy and retire a seven-year-old abandoned Thoroughbred in February 2004.

There are now more than two hundred equine rescue programs, from the unique Thoroughbred Retirement Foundation, which uses horses to help rehabilitate prisoners, to the more recent ReRun, which tries to find second careers for horses and raises funds partly by auctioning off paintings—called Moneighs—actually done by horses.

The programs are as new as the National Horse Protection Coalition, formed in 2003 and headed by Thoroughbred trainer Nick Zito, whose wife, Kim, has been retraining horses off Thoroughbred tracks for fifteen years. "You have to stand up and fight sometimes," he said. "I just don't want it to get ugly, because the horse is a beautiful thing. This is a creature of God, a horse. To kill them makes no sense."

Kim Zito, whose background is in show horses, intervened when she heard that her husband's first major stakes winner, Ride Sally, was destined for a sale that could have led her to a slaughterhouse. Now, when Ride Sally's second career as a broodmare is over, the owner will call the Zitos, who will find a home for her. "I think the biggest problem is ignorance," Kim Zito said. "People don't know. They don't realize that killers go to all the horse auctions."

Diana Pikulski knows. "I talk to shocked people every day," she said. "For so many years, people didn't realize what happens at auctions."

At auctions in places like New Holland, Pennsylvania, people look to make a quick buck by being middlemen in sending horses to slaughter.

But they cannot do that in California. On November 3, 1998, California became the first—and remains the only—state to explicitly prohibit slaughter, thanks to a referendum passed overwhelmingly. Some 59 percent of the 6.7 million ballots cast that day made it a felony to sell or export horses that will be slaughtered for human consumption.

Christine Berry of the Equine Protection Network said:

> The horse is ingrained in American culture as our partner, friend, and companion—not as our next meal. During the last twenty-five years, over four million of our beloved American horses have been secretly purchased and cruelly slaughtered to satisfy a foreign taste for horsemeat.

She calls the slaughter of America's horses an American tragedy.

Yet it still happens. Whether a horse winds up in a slaughterhouse by the direct action of an irresponsible owner looking for a few bucks or by the misrepresentation of killer buyers at auctions does not matter. The horse is born into the care of people, spends its life serving the needs of those individuals, and deserves a better fate than being carted off to a slaughterhouse in an overcrowded double-decker trailer built for cattle and then struck in the forehead with a captive bolt gun.

Susan Wagner, president and founder of Equine Advocates, calls horse slaughter "an American disgrace," and terms the annual slaughter of tens of thousands of American horses in the United States and Canada "the ultimate betrayal." The mission of Equine Advocates is "to rescue, protect, and prevent the abuse of horses through education, investigation, rescue operations, and the dissemination of information to the public."

According to Equine Advocates, 93 percent of callers in a 1995 national call-in television poll demanded the killing of horses for meat be banned. A 1997 poll in California showed 88 percent of California residents opposed horse slaughter. In 1999, a study by the New York State Senate showed that

91 percent of respondents considered horses companions, recreational, or sporting animals; 86 percent opposed the practice of slaughter; 73 percent believed that the manner horses are slaughtered is cruel and inhumane; and 72 percent would never eat horsemeat.

Most recently, a 2003 poll done by Mason-Dixon Polling & Research Inc. of Washington, D.C., revealed that 89 percent of Texas voters were unaware that horses were being slaughtered in their own state for human consumption overseas. Seventy-two percent said they were opposed to the slaughter of horses for human consumption. A May 18–20, 2004, poll by The Responsible Animal Care Society showed that 64 percent of Canadians did not believe horses should be slaughtered for human consumption.

For many, the slaughter of horses cannot be outlawed fast enough. "When they say that stopping slaughter is going to lead to mass neglect, that's a cop-out," Pikulski said.

Skip Trimble of the Texas Humane Legislative Network said, "What really appalls me is that people say this is a humane treatment. That is simply untrue. The vast majority of these horses slaughtered are young horses, healthy horses."

That is reality.

CHAPTER 5

Changing Reality

Howard Nolan, a longtime Thoroughbred owner and breeder, a past president of the New York Thoroughbred Breeders, and the state senator from Albany County for twenty years, had a simple idea in 1982: matching prisoners cleared for outside work with discarded Thoroughbreds nobody wanted. His concept has been a brilliant blueprint for the Thoroughbred Retirement Foundation, whose humble beginnings in Wallkill, New York, have grown to encompass similar facilities in Maryland, Kentucky, and Florida, with new ones being planned in Iowa, Oklahoma, Oregon, South Carolina, and Texas.

"It was funny," Nolan said. "There was a group [headed by Thoroughbred Retirement Foundation founder Monique S. Koehler] who was trying to figure out how to get a home for retired horses. At that time, I had Blue Sky Farm near Goshen, New York, and it was right near the Wallkill prison, a couple miles away as the crow flies. They had a farm there. And I thought to myself, 'Wouldn't this be great if we could join with the state and find something to do for the prisoners?' They had the land. They had the built-in labor; i.e., the prisoners. They would learn about horse racing and gain a skill that would be marketable when they got out. So, it was a natural. It was a way to use the land, to develop the skills of prisoners, and, from the horsemen's point of view, to have a farm for retired horses. That's how it happened. It was a win-win situation, a win for the prison system and a win for the retired racehorse with no place to go. A lot of them were being sent to slaughterhouses."

Prisoners at the Wallkill Correctional Facility have helped rehabilitate horses for more than two decades.

After getting approval from the New York State Department of Corrections, the Thoroughbred Retirement Foundation went to work converting the dilapidated dairy farm on the Wallkill grounds into a horse farm. The project needed a leader, and the TRF tabbed Jim Trimble, a fifty-two-year-old native of nearby Pine Bush who had managed two Standardbred farms in the area.

"I was between jobs," Trimble said. "When I came on-site, there were rusted-out stanchions in the barn, manure in the barn, and rusted-out barbed wire."

After beginning as the project manager for converting the farm, he became and remains the vocational instructor for the Department of Corrections.

The first step was to lower the floor of the barn to accommodate horses, which are taller than cows. "We got the jackhammer out, bulldozed dirt, and started building stalls," he said. "Then, we built the paddocks. Luckily, the farm here had a small bulldozer that one of the inmates could run."

They improved the drainage, built fences, and welcomed their first Thoroughbred, Promised Road, in July 1984. "He was foaled in 1975 and had nine wins and won about thirty-nine thousand dollars," Trimble said. The horse's owner, New Jersey trainer Ron Gibson, became farm manager so Trimble could set up and maintain the program for prisoners.

Supervision at the horse farm is minimal. Besides Trimble, there is a corrections officer on hand, and that is it. Could it work?

"I ran into some attitudes, 'You're not going to get the inmates to work,' or, 'You're not going to get the inmates to care,' " Trimble said. "I didn't think I would have a hard time to get them to care about the horses. That was not a concern. When you get working around horses, you get attached to them. They get under your skin. Horses read body language better than any animal and they remember forever.

"I was surprised by how hard the inmates worked. It was very rewarding working with them. I came in with the attitude that I'd have to be a tough guy or a taskmaster and push the guys to get things done. That has been the furthest thing from the truth. All I have to do is suggest something. Not every guy did it, but enough did so that things got done."

To participate, prisoners must be cleared by the Department of Corrections for outside work. Working with horses is not for all of the prisoners. "Some of them don't want to be outside in the weather," Trimble said. "Others just don't want to work. Others can go into programs that pay a lot more. At the horse farm, they get ninety-five cents a day."

Trimble said participating prisoners have ranged in age from nineteen to sixty-nine. They are in prison for crimes ranging from purse snatching to murder. The only crime that precludes a prisoner from participating is a sex crime. "I try not to find out what they're in for," Trimble said. "I don't want to be biased."

The only incident in the program's first twenty years was when two inmates, who were being interviewed by a female reporter, were trying to show off and got into a fistfight. It was broken up quickly and the two inmates were removed from the program.

The horses that are put into the program are usually in bad shape. "Wallkill gets the cripples," Trimble said. "Not many of those are ridable again. Some adoptions are for pets in people's backyards. Some get sent to other TRF farms."

And some do not survive. "A couple horses came here that were too injured and had to be put down," Trimble said. "That's a sad thing. But we've had other horses that vets have said should be put down, but Diana [Pikulski, Thoroughbred Retirement Foundation's executive director] said, 'Let's take him to Wallkill.' "

In the spring of 2004, Wallkill was home to thirty-seven Thoroughbreds, though it has had as many as forty-five at one time. A planned addition to the existing fifty acres will allow more horses to be in the program in the future.

What has transpired in the Thoroughbred Retirement Foundation's first twenty years is nothing short of remarkable. The program at the Wallkill State Correctional Facility remains a testimony to the bond between horse and man, one unfettered by the physical and emotional condition that either or both may be in when they first meet.

For many of the inmates, working with horses forces them into role reversals. "Some of these city guys are petrified," Trimble said. "They freeze when they come out here. It's pretty funny; the bigger the guy, the more scared they are. I think they're used to intimidating people because of their size, and now they're around bigger animals."

John Smith (the names of prisoners have been changed) was not a big, tough guy, but a pathetic thirty-year-old heroin addict facing four felonies. "Serious drug problem," Trimble said. "Emaciated. He was in rough shape. Working with the horses has saved his life. He's now a drug counselor with the Department of Corrections."

Trimble described Smith as apprehensive on his first day with horses. though "Not as scared as some of the guys," Trimble said. At first, Smith did nothing but watch. "He watched for two days," Trimble said. "He wouldn't go near the horses. Two days later, he started putting out feed and water. Within a week, he was leading horses in and out of the barn. After the first week, week and a half, he started grooming horses. He showed an interest in something he had never shown before."

Charles Jones was completely different. He was doing time at Sing Sing Correctional Facility in Ossining, New York, when George McGrath, now the deputy superintendent for the program, intervened. "He was not eligible for outside work," Trimble said. "Mr. McGrath saw him in the gym at Sing Sing. He said this guy is a perfect candidate to work for horses. He was a tough guy. He intimidates other people. Had a very negative view of life, not a fun individual to be around. Mr. McGrath said, 'If the horses are going to help anybody, this is the guy, to see if the program can make a difference.' This was maybe seven or eight years ago. We got him outside clearance."

When Jones arrived at Wallkill, "He didn't want to be told anything by anyone," Trimble said. Trimble placed Jones with one of the older inmates who was taking care of two horses that needed some extra work. After two weeks, Trimble allowed Jones to take care of the horses by himself.

"I spent a lot of time with him, hosing legs, picking feet, routine things," Trimble said. "And he accepted the challenge. After about two months he started talking to me: 'How do you do this? How do you do that? Can I have more horses?' "

Trimble put Jones in a field with more horses. "He was totally nonsocial with the other guys out here when he started. After a couple of months, he started talking to the guys. He became more socialized. And he's now a counselor."

John Adams had done two tours of duty as an Army Ranger in Vietnam before he was convicted for killing a policeman. He pleaded down to manslaughter and wound up at Wallkill. "He was a real tough guy, and he acted like that," Trimble said. "Everyone around him moved out of the way. He was coming through." After one week, Trimble matched Adams with Three Fires. "He was a very gentle horse I had worked with," Trimble said. "He was a nice horse."

One morning, Trimble was working in a nearby stall when he heard a loud cry of "Help! Help!"

"I go around the corner and see Mr. Adams is upside down with his head in the shavings on the floor and his back up against the wall, kind of crumpled up in the corner," Trimble said. "I looked and said, 'What happened?' He said, 'I guess I got a little aggressive with him.' He said the horse had kicked him, but he hadn't. He was holding the horse's rear leg and the horse kicked out, just kind of flicked him away. After that, this individual was not the tough guy. He did anything he could to help me. Did a lot of carpentry work. Big turnaround, 180 degrees, in his attitude."

Another inmate, who was six feet and 300 pounds, had been convicted for kidnapping and torturing a man who had not come up with the money he owed for drugs. "This guy was the enforcer," Trimble said. "He turned into a pussycat working around these animals. He really mellowed out."

Why?

"My theory on this, and it's only a theory, is that horses don't respond well when the person working with them is too tough or aggressive," Trimble said. "A person has to use body language to open socialization. They have to modify their behavior to get the most out of the horse. Instead of being a tough guy, they have to learn how to work with the horse. The slower they move their hands, the more they get out of the horse."

Thoroughbred Retirement Foundation's Pikulski is not surprised that working with horses can change a man. Said Pikulski:

> Everyone knows how working around horses feels. This could be a healing process for people who have no direction, people who have suffered from abuse in their lives, people who have suffered a lot. Horses can do miraculous things for the human spirit. That's why it works in prisons, in juvenile detention centers, in handicapped riding programs, and just with kids in the local high school here where I live in Vermont.
>
> Certainly, with the inmates, they have a lot of things going on. They lack empathy, self-esteem, and job skills. Every one of those things is addressed by being responsible in taking care of a horse. There's the added benefit that many of the horses who come to us are in real bad shape. They've been mistreated. They've been neglected. These guys are responsible for making them better. It's a universal healing thing. It's amazing.

Pikulski was teaching riding lessons while she was in high school in Middletown, New Jersey, when she met Monique Koehler, the founder of the Thoroughbred Retirement Foundation.

"She had just found out about the plight of these horses, and I helped her in the summers while I was in college at Drew University," Pikulski said. She continued to work for the foundation as a volunteer while she completed her education at Vermont Law School.

Following college, Pikulski served on the Thoroughbred Retirement Foundation board of directors until 1997, when the late Paul Mellon, a world-renowned philanthropist and owner of the prominent racing and breeding operation Rokeby Farm in Virginia, began to make significant donations to the foundation to fund the organization's public awareness campaign. The Thoroughbred Retirement Foundation was ready to grow. "They needed someone full-time, and I started January 1, 1997," Pikulski said. With donations continuing from his estate after he died, Mellon ultimately contributed more than $5 million to the foundation.

Under Pikulski's direction, the Thoroughbred Retirement Foundation has grown to accommodate 670 horses at farms in Connecticut, Florida, Kentucky, Maryland, Missouri, New Hamp-shire, New Jersey, New York, Okla-homa, South Carolina, Virginia, and Wisconsin. Besides Wallkill, there are prison horse farms at Marion Correctional Institute in Ocala, Florida; the Blackburn Correctional Complex in Lexing-ton, Kentucky; and the Charles H. Hickey Jr. School for at-risk juveniles in Baltimore, Maryland, with new ones planned in five other states.

In the spring of 2004, ground was broken for a Thoroughbred Retirement Foundation farm near the Wateree River Correctional Facility in Rembert, South Carolina. The impetus was a story about the foundation in the *New York*

Times that caught the attention of John Carmichael, then the warden of Wateree River.

Under Diana Pikulski, the TRF has grown to accommodate almost 700 horses across the country.

"The effect animals have on people is amazing, no matter where you are," Carmichael said. "On inmates, it has tremendous benefits. One of the biggest problems in South Carolina, which has become very much a Thoroughbred state, is that we don't have trained workers to work on our horse farms."

They will soon. And Gary Maynard, who was head of South Carolina's Department of Corrections and a proponent of the Thoroughbred Retirement Foundation farm at Wateree River, took a similar job in Iowa, and hopes to bring a Thoroughbred Retirement Foundation farm there.

There will never be a shortage of Thoroughbreds needing a home. On the backstretches of America's racetracks, there are notices reading:

The notice includes a retirement hotline, telephone number, fax number, and e-mail address. And as long as any Thoroughbred who has raced is free of contagious and communicable diseases and has no other retirement options, the horse is assured of a lifetime haven with the Thoroughbred Retirement Foundation.

Pikulski wants to do more. "What I'm trying to do is make this part of the racing industry's institutional makeup," she said. "In a number of ways, at various points in a horse's racing career, money changes hands. Are we going to look at all those places and put in a retirement fund, and, if we do, how do we distribute it? I think it should be on a horse-by-horse basis. Then, we are going to go a step further and police those

Do you know a horse in need of retirement? Thoroughbred Retirement Foundation offers a fast, first-rate alternative to the auction. No horses are sold. Transfer of your horse is tax-deductible.

places for people that are not inclined to do the right thing with their horses. You can take it a step further and say, are we going to change that bottom level of racing, which means horses getting out while they are still capable of having a second career?"

The foundation already has begun that process at the Exceller Farms, two farms donated by John Hettinger: a seventy-nine-acre farm in Poughquag, some forty miles east of Wallkill that opened in May 2001, and a thirty-acre tract opened in 2002 in Pawling, near the Connecticut border, which is an annex to Hettinger's own Akindale Farm, a Thoroughbred racing and breeding operation.

Michele Oren, a native of Bethel, Vermont, who retrains horses, runs Exceller Farms. Since its opening, more than one hundred fifty Thoroughbreds have been retrained for second careers and adopted, including three to mounted police units and seven to therapeutic riding programs.

"Racehorses are taught to run in a straight line," Oren said. "When they arrive here, the first step is to turn them out to pasture while they learn just to be horses. Then, we teach them things like how to turn and to stop. Every horse is treated as an individual."

The first horse adopted from Exceller Farm was Volcanic Ash, a then-four-year-old filly who was wary of any human contact. Judy Nelson, a young TRF volunteer who works at Akindale Farm, sat with her next to the grain bucket, day after day, and Volcanic Ash, who would eat only by approaching the young woman, slowly learned to trust people.

The Thoroughbred Retirement Foundation began a new adoption program at the Thoroughbred Retirement Foundation Secretariat Center at the Kentucky Horse Park outside Lexington in June 2004. It was conceived by Reiley McDonald, co-owner of Eaton Sales, one of the largest consignors of Thoroughbred yearlings, weanlings (horses less than a year old), and broodmares to public auctions in America. The Secretariat Center will accommodate as many as twenty horses on its twenty-two acres and showcase adoptable horses to an equine-friendly audience.

Keeneland Association, Thoroughbred owner-breeder Marylou Whitney, and Dr. Tony Ryan were major contributors to the Secretariat Center, which will be managed by the Thoroughbred Retirement Foundation but is open to horses from all equine retirement programs.

Whitney personally became involved in horse rescue when she learned, less than a month after her homebred Birdstone won the million-dollar Travers Stakes at Saratoga Race Course on August 28, 2004, that Birdstone's half brother, Cviano, had nearly wound up in a slaughterhouse. Named for Whitney's late husband C. V. Whitney, Cviano had won four of forty-nine starts and earned $73,883 on the track. The horse had been claimed from Whitney's stable three years earlier and the gelding had last raced for a $3,500 claiming price at Penn National in the spring of 2004.

In a story in *Thoroughbred Times* on September 25, 2004, executive news editor Michele MacDonald documented how two women, Beverly Strauss and Jo Deibel, of Angel Acres Horse Haven, purchased Cviano for $325 at the New Holland sale in Pennsylvania to save him from slaughter. When Deibel discovered Cviano's royal breeding, she contacted Whitney, who purchased Cviano for $2,500 and sent him to her farm in Lexington, Kentucky, to be used as riding horse.

"Thank God for these small organizations that are saving lives," Whitney's husband, John Hendrickson, said.

CHAPTER 6

California Leads the Way

The seeds of change can sprout from a single person standing up and saying, "This is wrong!" A woman in California did just that, and has saved the lives of hundreds, if not thousands, of horses headed for slaughter.

Cathleen Doyle, a Washington, D.C., native who lives in Sun Valley, California, with about a dozen horses, began attending horse sales in California in 1991.

"I went to a Barretts sale [in Pomona] and noticed that a lot of horses went to slaughter, especially broodmares," Doyle said. "Killers would buy twenty to twenty-five horses each time. You'd see the double-deckers coming in to get them. I started going to lower-end auctions on Friday and Saturday nights. They were worse."

Doyle knew firsthand that there is a distinction in California between horses and livestock, specifically cattle. "I knew when I went to the food store and bought a bale of hay, I had to pay state tax. The cattlemen who bought the same bale of hay didn't have to pay state tax. If we tax this animal as nonfood, then how can we be slaughtering them? If you go to a horse auction, you pay sales tax. At a cattle auction, you don't. There's a very clear line between food and nonfood animals.

"In our culture, we don't eat horses. Something is very wrong. And that kind of started me on my journey in 1991. I threw the gauntlet down and said we have to get an initiative to stop horse slaughter in California. But I had no idea of how to do it."

Others knew enough to help her find a way.

In late 1996, Doyle sent an information package to Joe Harper, the president and general manager of Del Mar

Thoroughbred Club, the scenic racetrack twenty miles north of San Diego and host to some of the best racing in the country. Harper, who in March 2003 was elected president of the racetrack industry trade group the Thoroughbred Racing Associations, is the grandson of legendary filmmaker Cecil B. DeMille and serves as trustee of the Cecil B. DeMille Foundation, which promotes higher education, child welfare, and film in Southern California.

Doyle, who founded California Equine Council to save horses, was not sure how Harper, or other racing executives, would react. "Racing was a huge contributor to slaughter," Doyle said. "I wasn't sure how they would feel."

But she was happy to learn that Harper was strongly opposed to slaughter. "He said he hated what he saw. He said he was sick to death of horse slaughter and, at the very least, they deserve a painless, humane death."

Harper ran ads for Doyle in the Del Mar track program calling for an end to horse slaughter. The ads read: *Save the Horses—Help Us Win This Race.*

"That was heresy at the time," Doyle said. "There was no talk of second careers for horses then. It was just breed them, trash them, and can them."

To get legislation passed to ban horse slaughter in California, Doyle formed a political action committee called Save the Horses. She needed 680,000 signatures to get a referendum on the ballot, and, after receiving Harper's support, she got 740,000.

Doyle found another ally in the racing industry in John Van de Kamp, then president of the Thoroughbred Owners of California and a member of the board of directors of the National Thoroughbred Racing Association. "He was a strong supporter and advocate of stopping the slaughter," Doyle said.

Van de Kamp said he supported Doyle's effort because of "concern over the cruelty involved in the present slaughter situation and the fact that we want to advocate more responsible treatment for horses we own and control."

Thoroughbred owners Bob and Beverly Lewis, who won the Kentucky Derby and Preakness Stakes with both Silver Charm (1997) and Charismatic (1999), offered their support to Save the Horses.

The organization was buoyed by a state poll conducted on March 8–11, 1997. Sponsored by Doyle's California Equine Council and conducted by Decision Research, the poll showed that 60 percent of Californians did not know that horses were slaughtered in America for human consumption. Sixty-nine percent said horsemeat should not be allowed to be exported from the United States, and 87 percent indicated they would not eat horsemeat if it were legal to do so in California. Thirty-eight percent considered horses as recreation and sporting animals; 31 percent thought of horses as companion animals; 23 percent saw horses as farm animals; 6 percent were unsure; and only 3 percent said that horses are a food animal to be used for meat, like pigs, cattle, and sheep.

According to the California Department of Agriculture, 3,112 horses were shipped out of state for slaughter for human consumption in 1997. In 1998, Save the Horses successfully got the initiative on the November ballot in California, as Proposition 6. Doyle was a cosponsor, along with Sherry DeBoer, Sidney Long, the late Tina Long, and the late Sue Maloney-Styles. Jeri Lerner handled the press conferences and ad campaigns.

"Our entire group had the fire in the belly, which one needs to win," Doyle said. "This was life and death. The horses' lives hung in the balance. Losing was not an option!"

They ran ads that read:

END HORSE SLAUGHTER

Dear Friends,

Every two minutes the animal that developed this nation in partnership with man … the very symbol of strength, beauty and America itself is slaughtered for human consumption in foreign countries … electrocuted into the kill box where they shake violently, urinating on themselves, betrayed, confused, terrified.

Please join California Equine Council and become an active part of this historic grass-roots effort. Together, we can Save the American Horse.

For the horses,
Cathleen Doyle

Van de Kamp, Doyle, and Priscilla Clark, who now runs her own rescue/retrain/retirement program at Tranquility Farm in California, met with editorial boards at newspapers to seek support for the cause, but they received little. "All but two papers, the *Los Angeles Daily News* and the *Bakersfield Californian*, opposed us," Doyle said. "The TV stations wouldn't run our ads. They said they were too graphic."

Doyle was not exaggerating. A report about Proposition 6 on Channel 4 (NBC) in Los Angeles noted that the ban slaughter ads "were deemed unfit for broadcast by every TV station in Los Angeles."

The opposition was considerable and backed by a white paper drafted by the American Horse Council, the American Association of Equine Practitioners, the American Quarter Horse Association, and the Professional Rodeo Cowboys Association that said horses are livestock and should have "salvage value." The white paper was ammunition for Proposition 6 opponents who were afraid the proposition was the first step to getting the United States Department of Agriculture to reclassify horses from agricultural animals to companion animals like dogs and cats.

John Harris, president of the Cattlemen's Beef Association and chairman of the California Horse Racing Board, led the opposition. On July 9, 1998, Harris sent a letter to the board of directors of the Thoroughbred Owners of California chastising them for supporting the ban on slaughter. He called the slaughter of horses humane.

Here is a graphic example of how "humane" the slaughter of horses is. The quote below is sworn testimony given in the state of Illinois in the mid-1990s, from a horse slaughterhouse employee, during a federal grand jury investigation of drugs in horses that were being sent to slaughter:

> I worked in the gut room ... when they bring the horses into the knock box they chain them up by both legs. Before they hang them up they remove the intestines through the butt. Sometimes the horses would still be shaking and wiggling. The horses were still living ... To get them into the kill box you had to shock them ... I was up there when he was hitting them with a baseball bat ...
>
> When we killed a pregnant mare, we would take the gut out and I would take the bag out and open it and cut the cord and put it in the trash and sometimes the baby would still be living and its heart would be beating, but we would put it into the trash can.

Basically, the pro-slaughter position commonly circulated contained these arguments against Proposition 6:

- This is not about humane treatment of horses. The initiative infringes on the personal freedom that all people currently have to choose what they want to eat—whether that choice is for religious reasons, for health reasons, for economic reasons, or for personal enjoyment.
- It is absurd that the proponents have titled this initiative "Save the Horses." The initiative does not save horses from being processed

for pet food; it does not save horses from blindness, disease, or injury; it does not save horses from starving to death due to cruel treatments by horse owners; it does not save unwanted horses from being abandoned by owners who could no longer take them to auction under the terms of the initiative.

- The initiative will result in a black market for horses going to slaughter.
- There are no horse slaughter facilities in the state of California; all are located in other states. If proponents are truly concerned about humane treatment of horses at slaughter facilities, they should be working to bring about changes in the laws in other states—not California. The initiative cannot change how other states treat horses.
- By failing to include provisions in the initiative that provide safer and more humane transportation of horses, and centering only on horsemeat consumed by humans, there can be no mistake about the proponents' real motive—to set a precedent for single-interest groups to dictate what humans can and cannot eat.

Another pro-slaughter faction, the Horse Industry Alliance, said the measure to ban slaughter was unenforceable for three reasons: (1) It is unconstitutional because it interferes with interstate commerce of livestock; (2) no agency exists with enough manpower to follow all vehicles hauling horses from California to determine their destination; and (3) the paperwork to ensure every horse sold out of California does not go to slaughter is overwhelming and unrealistic. Horse Industry Alliance officials said that Proposition 6 would negatively affect the welfare of "backyard" horses because their owners may not have the resources to pay a veterinarian to euthanize them.

But Jim Real, vice president of the California State Horsemen's Association, an affiliate of the American Horse Council, rallied support for Save the Horses. And celebrities, including Paul McCartney and his late wife, Linda, Robert Redford, Peter Falk, and Stephanie Powers, lent their support to Proposition 6. But Doyle is still bitter about the lack of help she received from others, including animal rights groups.

"They were nowhere to be seen," Doyle said. "Doris Day League is based in California and they didn't give one dime or do anything for the campaign. Neither did the HSUS [Humane Society of the United States], ASPCA [American Society for the Prevention of Cruelty to Animals], or PETA [People for the Ethical Treatment of Animals]. So all these animal rights groups now jumping on the back of our success in

California now makes me sick. It was all volunteer, all grassroots. No salaries, no offices, no expense accounts, nothing for a year and a half. You are just frazzled at the end. I walked into the end of this saying, 'We're going to lose this.' "

On election night, Doyle and ten of her principal supporters gathered in a room at the Biltmore Hotel in Los Angeles. "We didn't invite anyone because we thought we lost," Doyle said. "I was so tired and so burned out. We're sitting there in the Biltmore getting drunk. They weren't covering it on TV. And then someone walked in and said, 'Congratulations! You won!' I said, 'Oh my God, we did it.' "

It was not a photo finish. Fifty-nine percent of the 6.7 million voters chose to end horse slaughter for human consumption and the sale of horsemeat for human consumption in the state of California. The enacting legislation, which was added to the penal code of the state of California, mandates, "It is unlawful for any person to possess, import into, or export from the state, or to sell, buy, give away, hold, or accept any horse with the intent of killing, or killing, or having another person kill that horse, where that person knows or should have known that any part of that horse will be used for human consumption."

The penalty for breaking the law is a felony punishable by imprisonment from sixteen months to three years. Additionally, the legislation makes it illegal to sell horsemeat in the state of California for human consumption. Violators face a fine of $1,000 and/or a jail sentence of thirty days to two years. A second offense carries with it a prison term of two to five years.

"The smartest thing we did was to take the issue out of the hands of the horse industry and put it in the hands of American voters," Doyle said. "They're appalled."

Doyle has remained active in advocating the ban of horse slaughter throughout the United States through HoofPAC, the political action committee she began after Proposition 6 passed. And, more than three years after it passed, there has been no reported increase in cases of neglect. Meanwhile, some two dozen California programs try to put meaning into the victory of Proposition 6 by retraining, adopting out, or retiring the horses spared from slaughter.

CERF City

Grace Belcuore, a retired schoolteacher, wanted a horse when she was growing up in Lynn, Massachusetts. "I always wanted one, a young girl's dream," Belcuore said. "And I was never in a position to have them."

She has lots of them now. After moving to California, where she taught school, she began attending the races in the 1970s. She was one of thousands, if not millions, of fans who fell in love with Sam Rubin's great gelding John Henry, the Horse of the Year in 1981 at age six and the 1984 Horse of the Year at the advanced age of nine. John Henry won thirty-nine of eighty-three lifetime starts, earned a then-record $6,591,860, and was never at risk of winding up in a slaughterhouse. When he retired from the track, he went to the Kentucky Horse Park in Lexington, where he was seen by thousands of people every month.

But Belcuore wondered about the other geldings that were not as fortunate. "Here we're saving one guy, but what about the other guys?" Belcuore asked. "I was just devastated that these wonderful geldings . . . They make up the race cards. It really bothered me to think that we as a human race took everything out of the horses, and when we were done taking, we discarded them. And then, the journey and the death aren't very pretty. The whole scenario is devastating. And I still think it is. That's why I started CERF."

CERF, the California Equine Retirement Foundation, was incorporated in September 1986. It was granted tax-exempt status in May 1987, and found a home in Winchester, in Southern California, in 1988. Its goal was simple: "Give former racehorses a place to retire or help them find a new job and a new life."

In 1997, the California Equine Retirement Foundation's original mission was amended to include assistance to performance horses and an educational program for potential industry employees.

Since its inception, the California Equine Retirement Foundation has saved more than two hundred horses and currently has a horse population of seventy to seventy-five. Their lives were saved by this remarkable woman who said she was in her sixties when she first learned to ride.

None of her horses are dearer to her than Bedouin, the now-white son of Al Hattab out of Lady in Red, by Prince John. Foaled in 1981, the popular come-from-behind Bedouin raced in the 1984 Kentucky Derby, finishing fifteenth in the field of twenty in the race won by Swale. Bedouin finished his career with fourteen victories from sixty-seven starts and earnings of $311,394.

ReRun's Lori Neagle and CERF's Grace Belcuore help auction Moneighs.

"He's our mascot; he's beautiful," Belcuore said. "He amazed me, because when I got him, you could barely work with him. He'd side kick. He was a handful. Yet he turned out to be masterful with children. He goes to Del Mar every year and has kids all over him for America's Day at the Races or Children's Day, where the kids can pet him. He loves it. It's amazing because this was a tough guy to handle."

The California Equine Retirement Foundation got Bedouin in February 1988. "We took him off the track at Santa Anita. He came to us with a sinus infection and a swollen face. Apparently, he had an infected tooth. We had to lance and keep draining it. Dr. Greg Ferraro [D.V.M], who is now a director at Cal–Davis [the University of California at Davis], offered to operate on Bedouin for nothing if the horse could come to CERF. Dr. Ferraro was one of my mentors. He helped me get started, and he's one heck of an orthopedic surgeon."

Many others have helped the California Equine Retirement Foundation, including the late trainer John Russell, who was a board member of CERF and Tranquility Farm. Russell trained champions Susan's Girl and Track Robbery and numerous other stakes winners, including Cool Mood, Effervescing, Intrepid Hero,

Majestic Light, and Tri Jet. After retiring in 1995, Russell pursued a second career as a freelance writer and authored a novel. Russell died in 2004, at age sixty-seven, succumbing to stomach cancer at his home in Del Mar, California.

Other supporters of CERF include Del Mar and Santa Anita Park announcer Trevor Denman and jockeys Alex Solis and Gary Stevens. An honorary board includes owner Bob Lewis and celebrities Tim Conway, Dick Van Patten, John Forsythe, and Stephanie Powers.

But the star is Belcuore, who dared change the status quo of discarding racehorses once they could no longer produce on the track.

Charlie Brown

In the *Peanuts* cartoon, the character of Charlie Brown needed assistance anywhere he could find it, even resorting to paying Lucy five cents for psychiatric help. The Thoroughbred Charlie Brown had his life saved for not much more than a nickel by Helen Meredith, the founder and president of United Pegasus Foundation.

Meredith, who was born in Scotland and rode and trained in England and France before moving to the United States in 1989, began the not-for-profit United Pegasus Foundation, a 501(c)(3) charity, in 1994. Its mission is to "identify abused and/or neglected equines, aid and rehabilitate said equines, facilitate the adoption of said equines, and educate the public regarding the need to help these horses. The United Pegasus Foundation's goal is that no Thoroughbred should be denied a second career or a dignified retirement."

Meredith encountered Charlie Brown at a sale where he was odds-on to end up in a Texas slaughterhouse. The five-year-old horse was unbroken and suffering from malnutrition and past abuse to his face when Meredith saved his life. He had lost sight in one eye and paced endlessly in his stall after Meredith took him to the United Pegasus Foundation farm. After two years of retraining, he was adopted as an event horse by Alexis Young. In his first competition in a two-day event in October 2003, Charlie Brown finished second. It was another success story for Meredith and her husband, Derek, who trained Cardmania to win the one-million-dollar Breeders' Cup Sprint in 1993 and be named champion sprinter.

United Pegasus Foundation maintains two farms with a capacity of one hundred horses. Helen Meredith says the Thoroughbred industry should be doing more to help horses after their racing careers are over, especially when their careers are ended by injury.

In recent years, much of the United Pegasus Foundation's attention has been focused on saving mares and foals from PMU programs (in which pregnant mare urine is used to manufacture the drug Premarin). For more than sixty years, the urine of pregnant mares has been used to manufacture drugs that provide women with relief from symptoms of menopause.

"You never say to yourself, 'I've saved fifty horses,' " Meredith told Keith Morrison of NBC News in 2004. "You always say you had to leave fifty behind. It's so hard to think that they're all just going to go to a feedlot, be fattened up, and shipped for slaughter."

Gray Wins the White Horse

Leigh Gray is a veterinary technician for the Southern California Equine Foundation, which is a not-for-profit, charitable organization founded in the early 1980s. According to its website, it was founded by "a group of men who realized there was no existing organization whose sole function was to protect the interests and promote the welfare of the equine athlete."

Gray has not let eight cancer-related surgeries since childhood prevent her from helping horses. Gray, a former exercise rider, and her husband, Pete, have placed 112 injured Thoroughbreds through a program she began as an offshoot of the Southern California Equine Foundation. Her idea began when she asked her veterinarian bosses to perform pro bono surgeries on severely injured Thoroughbreds. They agreed, with the caveat that she would take care of them after surgery and find good homes for them.

"The best way to save a horse's life is to preserve his value," Gray said. "Trainers know I am available and that I will take the injured horse immediately. Because I work at the track, I can stop those horses from even starting down the road to killer buyers."

With the help of Santa Anita Park, which donated the land, and Oak Tree Racing Association, a not-for-profit organization that hosts a charitable fall meeting at Santa Anita and which put up 80 percent of the $400,000 cost, the Southern California Equine Foundation built an equine hospital on the backstretch at Santa Anita. Originally consisting of two mobile trailers, it now boasts a radiology unit and, on a grant left by philanthropist Dolly Green, a scintigraphy facility, which uncovers injuries not found on X-rays. With funding provided by Oak Tree, the foundation also developed a modern equine ambulance.

Leigh Gray dedicates most of her time to caring for horses and finding them new homes. She usually has at least five horses on her property at any given time and farms out others that she cannot personally house to friends and coworkers.

Among the horses she saved was Seria, a gelding so badly injured that veterinarians inserted a dozen screws into his damaged leg. He recovered under Gray's care to become a novice-level jumper. A similar success story happened with Trudy McCaffery's Attitude Approved, who recovered from a bowed tendon and has competed in several top-level riding events, including the Pan American Games, under her new trainer, Jill Watson. "Leigh knows how to match horses with the right owners," McCaffery said. "She places horses appropriately and always follows up."

But it was a single act of horsemanship that earned Gray the Race Track Chaplaincy of America's first White Horse Award and a check for $5,000 in 2003. She was working as the foot person on the horse-drawn carriage that takes the patrol judges to their stands at Santa Anita Park when fireworks caused the four horses to bolt. The driver was thrown from the carriage as it headed full speed toward the starting gate, which was filled with horses. Gray managed to stop the four runaway horses to prevent certain disaster.

Saving the Shark

The last thing Pam Berg needed in the spring of 2003 was another Thoroughbred at her Glen Ellen Vocational Academy farm in northern California. A former trainer and an associate steward who has worked with the California Horse Racing Board since 1988, Berg began Glen Ellen in 1995 to provide homes for injured, retired, and abused horses with the goal of adopting them out as either pleasure horses, show horses, or pets.

"All options are certainly preferable to the inhumane conditions of slaughter or needless euthanasia," Berg said. "Even 'unsound' horses can enjoy life and companionship, as do 'unsound' people. People are not euthanized or slaughtered because of an infirmity. So, why are horses subjected to such a fate?"

Certainly, the Thoroughbred named The Barking Shark deserved better. He had placed in a Grade 1 stakes, the Super Derby at Louisiana Downs in 1996, and earned $572,366 in his career, speeding to the lead in all his races and holding on for as long as he could. At age ten, he was still good enough to win a $5,000 claiming race at Golden Gate Fields on February 9, 2003. That gave him a career record of nine wins, thirteen seconds, and four thirds in fifty-six career starts despite racing with eight screws in one cannon bone. The Barking Shark was entered in an $8,000 claiming race at Golden Gate Fields for his next start. But owners-breeders Andy and Connie Pansini, Kit Hilling, Letha Ehrenfelt, and Teddy Cole, concerned about the possibility that The Barking Shark would break down from overextending his injured leg, stepped forward and bought The Barking Shark to retire him to Berg's farm.

Berg, a native of the Princeton area in New Jersey, has never been good at saying no to horses. She rode hunters and jumpers since her childhood. "I never got rid of the horse bug," Berg said. After college, she developed an import business specializing in horse tack. She kept riding horses when she moved to northern California in 1967, then was encouraged by a friend to breed Thoroughbreds. She became a trainer in the late 1970s, galloping her own horses. "I was not very good. I had my wins, but never had a big horse, never had a stakes horse." Her best horse was Sobranie, a filly who won a couple races on the northern California fair circuit in 1985.

In 1987, an accident changed her career and nearly took her life. She got trampled by a filly she was walking who was spooked by a swinging gate. Berg sustained a punctured lung, broken scapula, broken ribs, bruised heart, concussion, broken nose, and multiple injuries to her face. As she was recuperating, doctors discovered that she also had broken her shoulder. "I'm lucky to be alive," she said.

She was advised to curtail her activities, but she did not enjoy training horses without also galloping them. So she became a steward in 1988, and she took on a new mission.

"I realized the need for retirement facilities for horses. They needed a place to go. So many were needlessly euthanized because people didn't want to spend the time to lay them up and people didn't want to spend the money for costly surgery. If they weren't productive, just get rid of them. And the easiest way to get rid of them is slaughter."

Berg originally wanted Glen Ellen to serve as a retraining program for racetrack workers coming out of rehab for substance abuse by matching them with horses at her ranch. Funding for the program failed to materialize, so Berg has operated the farm as a not-for-profit home for horses. All but two of her twenty-eight horses are Thoroughbreds. She estimates that she has had eighty horses at the farm since Glen Ellen started in 1996, and she has placed many of them. "Half of my evenings are spent reading e-mails needing to place a horse," she said. "Occasionally, someone e-mails wanting to adopt one."

The Barking Shark was the twenty-eighth equine resident at her eight-and-a-half-acre Glen Ellen farm in Sonoma County in the heart of wine country. It was not a hard sell. "I knew before he was purchased that he was coming here," Berg said. "How can you turn a horse like that away?"

Pam Berg, center, receives the Dominion Award from Cot and Anne Campbell.

A half dozen volunteers help Berg operate Glen Ellen, but most of the work is done by her. Why? "I guess it's the same horse bug," she said. "There's a need and someone needed to do something about it."

On August 4, 2004, Pam Berg was recognized at Saratoga Race Course in New York for doing "something about it." That afternoon, at a luncheon hosted by Dogwood Stable, Berg was presented the twelfth annual Dogwood Dominion Award, named in honor of Dogwood's Thoroughbred stakes winner Dominion and given to those who have distinguished themselves in service to the Thoroughbred industry. Berg received a check for $5,000 and a bronze statue of Dominion.

"She feeds these horses, mucks all the stalls, repairs fences, and organizes fundraisers to help support her charges while commuting to her job as a steward at several northern California racetracks," Dogwood Stable President Cot Campbell said in admiration.

After the luncheon, Berg was escorted to Saratoga, America's most famous racetrack, where the fifth race had been named in her honor. She then presented the trophy to winning jockey Jerry Bailey, who had just ridden Daydreaming to victory. More people had learned about her work at Glen Ellen than she ever could have envisioned.

Tranquility at Last

Deep down, Priscilla Clark knew that her life's work would be with horses, even if it took her a little while to realize that. Growing up in Kalamazoo, Michigan, she began riding horses at age six. When she was twelve, she had the good fortune of being taken in by a mentor, Harriet Hayner, in nearby Galesberg.

"She was like my second mom," Clark said. "She taught me all there is to learn about riding and caring for horses. I worked with her from when I was twelve through college."

At Western Michigan University, Clark received a degree in philosophy and English. She married for the first time, had a daughter, Gwen, and still rode a pleasure horse. But she felt unfulfilled.

"I decided I just could not be happy doing anything but being a horse professional, even though there wasn't a lot of money in it," she said.

She went to Oregon, attended a horseshoeing school at Oregon State University, and became a certified farrier. Settling in northern California, where she practiced for twenty years, she married trainer Stan Waterman after meeting him at a small fair track. Clark began taking in horses that had been injured or sick and needed to be trained to get back to the racetrack. But some of the horses she rehabbed were not sound enough to return to racing.

"And the owners said, 'Good-bye, find a home for him,' " Clark said. "That's common. That's why they end up in slaughter. I found myself with all these racehorses who I got healthy but knew nothing about being a riding horse. To find them homes, I had to retrain them so someone else could use them."

She did that successfully, frequently working the horses on a nearby beach. "I'd canter them on the beach and walk them in the salt water," Clark said. Word quickly spread at the nearby racetracks. "People would send injured horses to me," Clark said. "People would come up to us and say, 'Will you take this one? Will you take that one?' We got overwhelmed. It just got to be way too much."

At one point, Clark was tending to forty horses. Then, she and her family relocated to central California in 1996, and she became acquainted with Gary Biszantz, the former chairman of Cobra Golf Inc. and who is a past director of the Thoroughbred Owners of California, chairman of the Thoroughbred Owners and Breeders Association, and on the board of Breeders' Cup Ltd. Biszantz was introduced to racing by his father, Harry, and has owned horses since 1956. His stakes winners include Old Trieste and Cobra King.

"Gary and I had a lot of phone conversations in 1996 and 1997, but we never met," Clark said. "His dad had passed away, and he indicated to me that if I could find a suitable farm that was affordable and functional and would work for retraining or retiring Thoroughbreds, he was interested in buying one in honor of his dad."

Clark began checking out available farms without much luck. "I kept going out and looking at farms, and they were a million dollars," she said. "We got very discouraged. It went on for two years. It got old. I thought this was never going to happen."

But it did. She spotted an ad in the *California Thoroughbred*, a monthly magazine published by the California Thoroughbred Breeders Association, for a forty-acre farm in Tehachapi, which is about 120 miles north of Los Angeles, with three barns and seventy-five stalls for $240,000. She thought the price was a typographical error, that it was really $1,240,000. When she called, she learned that the price for the farm, which needed a lot of work, was indeed $240,000. She called Biszantz and they met in person for the first time at the gate of the farm.

"We walked through the property," Clark said. "It was daunting, but there were three big barns and forty acres, level. Gary said, 'We're going to buy it.' "

They bought the land in the spring of 1998, creating the Harry A. Biszantz Memorial Center at Tranquility Farm, currently home to eighty-five Thoroughbreds.

"We had to do everything from scratch—outbuildings, fences, and pastures," Clark said. "Now it's all irrigated and cross fenced with seven pastures with big run-in sheds, a hay barn, a riding arena, my residence, and a covered round pen for training. There are big, green fields. It's really pretty.

"We have a very, very clear mission. We are a service to the Thoroughbred industry as a whole. We have a meticulously structured program to lead to as many adoptions as possible."

Tranquility Farm's board of directors, led by Clark and Biszantz, include owner and breeder Trudy McCaffery, former Mattel Chairman and Chief Executive Officer John Amerman, Eclipse Award-winning writer Jay Hovdey, trainer Jenine Sahadi, bloodstock specialist Rollin Baugh, and Santa Anita Park and Del Mar announcer Trevor Denman. The horses that are not adopted become pensioned in this lush setting. Of her eighty-five current horses, Clark said that twenty-five are in rehab and retraining. A handful of others are in long-term turnout and may be retrained.

Tranquility Farm in Tehachapi, California, is home to 85 rescued horses.

There are also five broodmares with foals. The rest are retired. Clark estimates she has averaged adopting out twenty-five horses a year since opening the doors of Tranquility Farm in March 1998.

Besides redirecting the lives of Thoroughbreds who otherwise would have gone to slaughter, Clark has become an eloquent spokesperson on the issue. She authored the "Final Turn" column in the February 28, 2004, issue of *The Blood-Horse* on the topic. Finding exception with the American Association of Equine Practitioners, which is lobbying against the American Horse Slaughter Protection Act by contending that death in a slaughterhouse is humane, Clark wrote:

> If the AAEP believes the slaughter of horses is humane, you may ask why the manner in which Ferdinand was put to death [in a slaughterhouse in Japan] was so appalling to the racing world. The answer is quite simple: the American public understands implicitly that humane euthanasia means death by lethal injection and not being bludgeoned into unconsciousness by a captive bolt in a slaughterhouse.
>
> The American public is not the least bit confused about the consequences of administering these two very different means of render-

> ing death. They understand that the horse, a highly intelligent animal, dies fighting for his life in the slaughterhouse instead of peacefully passing by tranquilizers administered in a stress-free environment. How the American Association of Equine Practitioners, you who are entrusted by the American public to protect the welfare of horses, could fail to make this distinction and equate death in a slaughterhouse with humane euthanasia is nothing short of remarkable.

When Proposition 6 passed in California, anti-slaughter advocates hoped it would spur other states into action. Through early 2004, it had not.

However, one state has had a slaughter prohibition on its books for more than fifty years, yet this state still allows slaughterhouses to kill horses for human consumption.

CHAPTER 7

Killing Them in Texas

Mary Nash and Paula Bacon have deep roots in Kaufman, Texas, a city with a population of 6,600 located thirty-three miles southeast of Dallas, where Belgian-owned Dallas Crown Inc. continues to slaughter horses for human consumption overseas.

Nash's great-grandfather, Charles Cornelius Nash, was driving a herd of horses across Texas looking for good grazing land when he settled in Kaufman and purchased a forty-six-acre farm, which is now run by Mary Nash. The farm is adjacent to Dallas Crown. "It was originally for cattle under another name," Nash said. "The plant was built to slaughter cattle, not horses."

Though Nash's dad worked in insurance and real estate, he loved horses. "My dad and I used to train Shetland ponies," she said. "I rode my horse around town like a lot of kids here. I grew up with horses."

After Nash graduated from Kaufman High School, she attended Southern Methodist University, got married, and lived in Dallas with her husband and their son, Joseph. The family moved back to Kaufman in 1987 when Joseph was three. She took a job working in computer technical support for an accounting firm until she retired in 1998.

"From 1990, I had unfettered access to that property," she said. "I've been seeing the horses down there [at Dallas Crown] for years. We lived within walking distance from the farm, a couple blocks away. We'd go over in a golf cart. I remember the first time I got a good look at those horses in December 1990. I was amazed."

She was amazed because she had spent her entire life believing that the horses who were slaughtered at Dallas Crown were horses who were either injured or diseased.

"I thought those horses look a lot better than any horse I ever rode when I was a kid in Kaufman, a lot better," she said. "I thought, 'My God, why are these horses being killed?' These are horses who are gorgeous. I was so disgusted. It made me so mad. I didn't think there was a thing I could do about it."

She remembers the stories she had heard growing up. "The stories floating around town were that they all had diseases," she said. "That's what they were telling people when they slaughtered horses, that it was okay to slaughter them. That was the story being passed around town. For years, a friend of mine who's a vet would tell the story about the horses' sicknesses." In 2003, one of Nash's neighbors was having a beer with Nash's husband, Stephen, a lawyer who works in Dallas. "[Our neighbor] said he was having a beer with an employee of Dallas Crown and the guy said Dallas Crown's biggest client is Disney World to feed their zoo animals," Mary Nash said. "I said, 'He was lying.' "

Then Nash went on the Internet, downloaded a couple articles about horse slaughter for human consumption overseas and left them in the Dallas Crown's employee's mailbox with her business card. "Never in my life did I have an easier time dispelling a lie," she said. "That's the kind of bull that goes on in Kaufman, where people are naïve and unsophisticated. They believe anything without questioning it."

She knows because she did—until she got a closer look at reality. She has watched underground video filmed by the Humane Farming Association of the actual slaughter of horses at Dallas Crown when the plant did not know a visitor was coming. Asked if horses are always killed with the first shot of a stun gun, she said, "Of course not. Haven't you seen the video? A cow has a short neck like a hog. They get them in confined spaces. I've been around cattle; my dad used to raise cattle. And I've been around horses. The horse has a long neck and horses get excited. They flail around. Cows don't do that. With a horse, they're jumping around like crazy. They're nervous. They're scared. How are they going to hit the right spot with them jumping around? They hit everything but the right spot. It's a complete betrayal [of the horses]. They trust us and work for us. And then we sell them to these Belgian jerks?"

The people who sell horses and the people who buy and slaughter horses are doing so contrary to the state law of Texas, a law in existence since 1949.

When Skip Trimble, a sixty-four-year-old lawyer who owns Thoroughbreds, was asked in 2001 to draft legislation that could possibly stop the slaughter of horses for human consumption in the state of Texas—where two of the three horse slaughterhouse plants in America are in operation—his research revealed an interesting fact: a statute banning horse slaughter for human consumption already existed.

According to Trimble:

> The law said it clearly in Chapter 149 [Sections 149.002 and 149.003] of the Texas Agricultural Code. That law basically says you can't sell, exhibit for sale, possess for sale, or transfer for sale horsemeat for human consumption. If you do, that's a crime, punishable [by] up to two years in jail. I said this is pretty strange because one of the plants has been doing it since the [nineteen] seventies.

The French-owned Beltex Corporation has been slaughtering horses for human consumption since 1978 at its plant in Fort Worth, roughly twenty miles west of Lone Star Park, the host of the 2004 Breeders' Cup World Thoroughbred Championship races, in Grand Prairie. The plant also sells bison and ostrich meat to markets in France, Belgium, and Japan. The plant has some ninety employees.

The Belgian-owned Dallas Crown Inc. in Kaufman, forty miles southeast of Lone Star Park, slaughters horses for human consumption in Italy, Japan, France, Belgium, Switzerland, and Mexico. Dallas Crown, which opened in 1994, has about forty employees.

In 2002, the two plants slaughtered more than 42,000 American horses and shipped the meat overseas, resulting in gross sales of more than $40 million.

In 2003, 50,564 American horses were slaughtered at the two Texas plants for human consumption, according to the United States Department of Agriculture.

According to Just Say Whoa!! To Horse Slaughter, approximately five thousand horses were slaughtered at the two plants in the month of January 2004. They were slaughtered despite the law prohibiting such a business operating in the state.

"When I saw that law in the books, I started calling the Department of Agriculture and district attorneys," said Trimble, who is on the legislative committee of the Texas Humane Legislation Network. "I asked, 'Why is it that these plants are still in operation?' Nobody seemed to know. So I started writing letters back and forth, asking for explanations, and never got anything satisfactory. I said, 'Why not ask the attorney general to give an opinion?' "

The Texas Attorney General, John Cornyn, who has since been elected to the United States Senate, issued his opinion on August 7, 2002. Discarding the companies' contention that they were not subject to Texas state law because the meat was sold overseas, Cornyn ruled that Texas state law did apply and that the owners of the companies should be prosecuted.

But the companies jointly filed a federal lawsuit in September 2002, and procured an injunction to remain in business until that lawsuit is adjudicated. The two Texas county district attorneys involved agreed to wait until the lawsuit was resolved before prosecuting. (In the suit, Beltex included its slaughterhouse, Empacadora de Carnes de Fresnillo S.A. de C.V. in north central Mexico, because the company ships horsemeat through Texas to be flown to foreign countries.) The lawsuit is awaiting a decision from U.S. District Court Judge Terry Means in the Fort Worth Division of the Northern District in Texas.

In the interim, Representative Betty Brown of Kaufman went to bat for the slaughterhouses, sponsoring a bill that would have amended the Texas state law banning horse slaughter only if the human consumption of horsemeat occurs in the United States. "We have a population of over a million horses in Texas and very few end up in slaughterhouses," Brown said. "But this is one avenue that needs to remain open."

The reality, of course, is that horses sent to slaughter at the two Texas plants come from all over the country and are frequently transported inhumanely to be slaughtered inhumanely.

Not all horses that end up in slaughterhouses were sent by owners who knew of their fate. Some end up there under tragic circumstances, their owners duped by people who are looking to make a quick buck. Judy Taylor of Kentucky knows.

According to documents from the Jefferson Circuit Court of the Commonwealth of Kentucky, two of Taylor's horses were slaughtered at Beltex after she had arranged for other people to care for them.

Taylor owned two registered Appaloosas named Poco and P. J. Poco was fourteen and P. J. thirteen, and Taylor had had them since they were foaled. When she and her husband separated in 1994, Taylor remained at the residence where the horses were boarded and assumed sole responsibility for their care. Due to deteriorating health—she was suffering from myasthenia gravis and other medical problems—she could not continue to physically take care of the horses. Instead of selling them, she entered into an arrangement called a "free-lease" agreement, in which someone would care for the horses in exchange for the enjoyment of having them.

Taylor's brother suggested she give her horses to his friends Lisa and Jeff Burgess, who had a small farm with horses of their own. Taylor testified that she explained to Lisa Burgess that she never wanted to lose contact with the two horses, and if the Burgesses could no longer care for them, they would be returned to Taylor. Taylor did not transfer ownership of the two horses.

On August 31, 1994, the Burgesses took the horses. Lisa called Taylor that evening to tell her that she had led them around their new pasture and that the horses were doing fine. Within a few days, Lisa Burgess called Eugene Jackson, a known killer buyer, to say that she had two horses for sale. On September 6, 1994, Jackson purchased the two horses for $1,000.

Taylor waited a week to visit her horses, bring them treats, and take pictures. At that point, Lisa Burgess told Taylor that she had given them to a man she had met on a trail ride, but she did not know his name. Upset and frightened, Taylor said she needed to know where her horses were so she could bring them back home.

The Burgesses then got a friend in Indiana, Kenny Rudolph, to cover for them by lying and telling Taylor that he had the horses. When Taylor called and said she wanted to see the horses, Randolph gave her vague directions to a fictitious location in Frenchtown, Indiana. After Taylor was unable to locate the horses, she returned home distraught. Randolph later confessed to an Indiana police detective that he had lied to Taylor and never had the horses.

With the aid of Victoria Coomber, a humane investigator, and Sharon Mayes, president of a local humane organization, Taylor learned that Eugene Jackson had bought the two horses from the Burgesses and sold them to Jason Ryan of the Ryan Horse Company, a business that supplies horses to Beltex to be slaughtered. They died there in late September 1994.

On August 23, 1995, Taylor filed suit against the Burgesses, Randolph, and Jackson in Jefferson Circuit Court in the Commonwealth of Kentucky. She filed an amended complaint to include Jason Ryan, James Ryan, and the Ryan Horse Company as defendants. The lawsuit against Randolph was dismissed for lack of jurisdiction, and that against Jackson and the Ryans was dismissed on grounds of improper venue.

In the suit against the Burgesses, a jury in April 1999 ruled that the couple had breached their agreement with Taylor and that they had intentionally inflicted emotional distress on her. The jury awarded Taylor $1,000 for the fair market value of the horses, $50,000 in compensatory damages for outrageous conduct, and $75,000 for punitive damages. The Burgesses lost on all appeals.

While Taylor's horses were sold by unscrupulous friends, others sold to slaughter in Texas are stolen. Texas Representative Charlie Howard said on the floor of the Texas House of Representatives on April 23, 2003—the day Representative Betty Brown's pro-slaughter bill was voted on—that he learned his stolen horses ended up in one of the slaughterhouses.

Brown's bill passed the Texas House of Representatives, 83–53, that day. The bill, however, was killed in the Senate despite an effort to resuscitate it by Representative Rick Hardcastle of Vernon, who attached an amendment to an omnibus agricultural bill sponsored by Senator Robert Duell of Greenville that would have allowed human consumption of horsemeat. Duell was not pleased at the addition and stripped it from his bill in a conference committee.

"They tried to get the law changed, and they failed at that," Trimble said. "So we're back to square one, which is, 'What is the judge going to say about the lawsuit?' "

In early March 2004, Ann Diamond, the Assistant District Attorney for Tarrant County, said the judge's decision could come in days or it could be months. At the time Diamond said this, the 2004 Breeders' Cup was less than eight months away,

"We hope that it will be over by then," Trimble said in early 2004. "I'm sure there will be huge demonstrations if the slaughterhouses remain open, but it won't be by our group. There are horse groups that are irate over this."

At the end of 2004, the two Texas slaughterhouses were still in operation.

Personally, Trimble is appalled. "I saw Ferdinand in the Kentucky Derby. I was there. I still get a lump in my throat when I think of him being slaughtered in Japan. That's the last time Willie Shoemaker [who was a native of Texas] won the Derby. I saw Ferdinand beat Alysheba in the Breeders' Cup Classic at Hollywood Park."

Trimble has impatiently waited for the federal court to rule on the lawsuit to end the slaughter of horses in Texas, but nothing has changed. "It's status quo," he said. "They're still slaughtering horses as fast as they ever did. And nobody can do anything."

But people are trying. Mary Nash began her own website after she was appointed to the Kaufman Planning and Zoning Commission in 1999. The following summer, she expanded the zoning website, www.kaufmanzoning.net, to include a section on horse slaughter. That section offers a timeline of important events since 1997, undercover scenes from a Texas horse slaughterhouse, the Texas lawsuit documents, the American Horse Slaughter Prevention Act, a message from Chris Heyde of the Society for Animal Protective Legislation, and an exhaustive list of correspondence on the topic.

Skip Trimble pressed for enforcement of a 1949 Texas law that bans horse slaughter for human consumption.

On March 22, 2004, a TV crew from Dallas aired a piece about Dallas Crown. To get a view of the horses at Dallas Crown before they were slaughtered, they shot film from Nash's farm adjacent to the plant. "The mayor and I were both in it," Nash said.

In response, Dallas Crown built a barrier between its property and Nash's to obstruct the view. "Nothing surprises me," Nash said. "These people are liars. People in Kaufman have been lied to for thirty years."

In a July 25, 2004, e-mail to Carol Chapman, who operates The Last Refuge in Hitchcock, Texas, Nash wrote:

> If you feel like giving up, come to Kaufman and look at the horses waiting for slaughter. I remember one day I drove to our farm on our golf cart to check out Dallas Crown to see how many horses were waiting.
>
> It was late afternoon and the killers had gone home, leaving the horses behind to mill around in the holding pen. When the horses saw me across the way, they perked up and watched me.
>
> I felt like they were expecting me to come over and talk to them. A few stood patiently at the fence, heads high, as if to say, 'Aren't you coming?'
>
> It broke my heart.

Paula Bacon, elected mayor of Kaufman in May 2003, is doing everything she can to close Dallas Crown. "It is gruesome, inhumane, and everything you think it would be," Bacon said in July 2004. "It has an incredible smell with thick, thick black flies.

"This morning in Texas, I got up at 6:30. I live in a house my grandfather bought. I open the back door, and then it hits me, the smell of urine and blood. They're killing today. They're killing on Mondays and Thursdays. I'm looking for fresh air. What I got was a nose-full of manure, urine, and either blood or terror. They smell death. They're terrified."

Bacon, who is a native of Kaufman, lived in New Mexico, Boston, and Italy before returning to Kaufman to help her dad with his lumberyard.

When she was growing up in Kaufman, she, too, believed the line that the slaughterhouse was a necessary evil. "People would go around saying these horses are old and sick and have a defect in their legs that would cripple them later in life," she said. "I remember stopping and thinking, 'This sounds stinky. Surely, there must be a regulatory commission with people more intelligent than I am on it. It must be okay.' As Mary [Nash] said, it slipped under the radar."

And that is where it has stayed for years.

"There's a mentality here that has to do with people in this situation," Bacon said. "They enter a denial phase. They don't want to deal with it. People have been lied to. People trusted the officials. Now, if you admit it's wrong, you have to admit that you were fooled for years and years and years. They can't admit horse slaughter was a big lie all these years. It's very frustrating.

"I'm looking for a lawyer to close those [expletives] down."

CHAPTER 8

Back in Business in Illinois

When the Belgian-owned Cavel International Inc. slaughterhouse in DeKalb, Illinois, burned to the ground in March 2002, amid unproven accusations that animal rights activists were responsible, Cavel did not wait long to announce it would rebuild the plant and reopen it as early as April 2004, so it could resume slaughtering as many as one hundred horses a day. It took a bit longer than that, but Cavel did, indeed, reopen in early June 2004.

By then, a rapidly growing base of opposition had sparked legislation in the Illinois Legislature, which would have explicitly banned the slaughter of horses for human consumption. Senator Christine Radogno of Lemont and Representative Robert S. Molaro of Chicago introduced companion bills in their respective houses to outlaw horse slaughter.

"When we were informed that there would be a plant opening for the slaughter of horses for human consumption, I would have to say almost every legislator in Springfield was shocked," Molaro said in an interview with Chicago's ABC Channel 7 in March 2004. "We don't breed horses to slaughter them for human consumption. That's what we're going to outlaw in this bill: not in this state, not in this country. And that's what the bill does. It's that simple."

Molaro had the ardent support of Arlington Park Chairman Richard Duchossois, whose track is located in the Chicago suburb of Arlington Heights. Said Duchossois, "It's the right thing to be done because it's the right thing to do."

But the 500-member Horsemen's Council of Illinois, which is affiliated with the American Horse Council, opposed the ban.

Thoroughbred trainer Nick Zito, spokesperson for the National Horse Protection Coalition, could not believe it. Said Zito:

> When I learned that the Illinois Horsemen's Council supports horse slaughter and says a ban is going to hurt our industry, I was taken aback. I have been a horseman all my life. It is what I do every day, and to see them claim to represent the 'horse industry,' you can imagine I was a bit surprised. Horse slaughter is a dirty mark on our industry, which is why support for the American Horse Slaughter Prevention Act and efforts to ban slaughter are growing. Banning horse slaughter is not only good for the industry, it is the right thing to do for the horses.

One of the Horsemen's Council spokespersons, R. D. "Dean" Scoggins, has served as equine extension veterinarian at the University of Illinois College of Veterinary Medicine for some twenty-seven years and is a member of the American Association of Equine Practitioners. In a November 2003 letter to the legislative committee considering the anti-slaughter bill, Scoggins wrote that he had been to the Cavel plant at least ten times and that the captive bolt method of killing horses "is considered by persons qualified in neurology and anesthesia as one of the most humane means of death available." He also said the captive bolt method was effective in immediately stunning a horse. "I have not seen a partially stunned horse," he wrote.

Steve Hindi, a former hunter and fisherman who started SHARK—Showing Animals Respect and Kindness—has made the reality of horse slaughter easy to see for Illinois legislators and anyone else interested in seeing the barbaric reality. On one-hundred-inch screens in front of the Springfield capitol building, Hindi has played videotape of a horse being slaughtered at one of the Texas plants. It was shot by an associate from the Humane Farming Association. The film is gruesome. "They are very intelligent, very aware animals," Hindi told Garrett Ordower in a February 29, 2004, story published in the *Daily Herald*, which is based in suburban Chicago. "They can feel death. They can smell blood." If they are lucky, they are killed with the first shot of the bolt gun. But not all of them are lucky.

Anyone with a constitution strong enough to watch can check out the video at SHARK's website, www.sharkonline.org. There, under Horse Slaughter Horror Video Documentation, one can see the video for free. For those who do not want to witness the savagery defended by

Horse Slaughter Horror Video Documentation

Narrative description of SHARK's horse slaughter video at www.sharkonline.org:

***

"Doomed horses are lined up for cruel termination. The last hours of these intelligent animals are filled with dread as they experience the noise, smell, and sight of death."

***

"This horse is hit in the shoulder by the misaimed captive bolt gun, causing even more pain and trauma.

"Wounded but still conscious after the misaimed shot, the suffering victim is hit again and again by the merciless captive bolt gun."

***

"A panicked victim struggles to gain footing on the slippery blood-soaked floor of the kill chute.

"Finally steady, the horse is shot by her killer, but she remains alive and conscious.

"Still struggling, she is shot a third time by her killer."

***

"Clearly aware of the peril she faces, this horse repeatedly attempts to avoid thecaptive bolt gun.

"The horror continues as the doomed horse continues to evade death. She looks toward her killer with eyes wide with terror. If she is looking for mercy, there is none to be found in this death house.

"Finally, her killer succeeds in blasting her skull, but even after three shots she is still moving."

***

"A still-moving victim is ejected from the kill chute.

"The latest victim is hoisted, to be cut up in view of still-living victims.

"The victim on the left side of the screen is having its face cut apart, while the victim on the right is being bled. This occurs in view of other horses awaiting the merciless captive bolt.

"While the dismantling of these once-beautiful creatures goes on, the camera pans to the kill chute just a few feet away. There a victim is hit four times with the captive bolt gun."

the American Association of Equine Practitioners and the American Veterinary Medical Association and the American Quarter Horse Association, there is a narrative description of the video (see box). Judy

Marteniuk, D.V.M., who is a faculty member at Michigan State University's College of Veterinary Medicine, where she specializes in equine medicine in the Department of Large Animal Clinical Sciences, co-produced the film.

But the website also shows six film clips of humane horse euthanasia, with the commentary: "These six clips add up to a total of one minute, and show two horses being put to sleep. They were euthanized due to failing health. They were put down together, as they knew each other, and were very close. Their final day was sunny and breezy, and it is obvious that they are completely calm and at ease. They are given loving care from the beginning to the very end. This footage demonstrates how people who truly care about their horses can give them a peaceful, dignified end. The horses simply become groggy and fall asleep. What a complete and utter difference from the horror of the slaughterhouse!"

"I find it disingenuous that its defenders claim horse slaughter is a form of humane euthanasia akin to veterinary-administered euthanasia," said Lydia Gray, D.V.M., the executive director of the Illinois-based Hooved Animal Humane Society. "The American Horse Slaughter Prevention Act presents a clear opportunity to end one form of cruelty, and anyone who truly cares about horses should embrace this opportunity."

On March 25, 2004, Molaro's bill to ban horse slaughter in Illinois passed the Assembly 55–54, but the measure needed a constitutional majority of sixty to move forward. However, a number of representatives who were supporting the bill missed the vote for various reasons, and a new vote was requested. But after the anti-slaughter bill passed the Illinois Senate, 38–15, on May 24, the bill was defeated four days later when the Assembly did vote again, voting against the anti-slaughter bill, 61–50. The bill's backers, including State Senator John Cullerton of Chicago, vowed to continue the fight.

Meanwhile, in New York, on May 3, 2004, Assemblywoman Deborah Glick of New York City introduced a bill in the New York State Assembly prohibiting horse slaughter and the selling, bartering, purchasing, possessing, transporting, delivering, or receiving of horse flesh intended for human consumption. "There aren't any slaughterhouses in New York and we want to ensure there won't be," Glick said.

CHAPTER

9

The Act

Passage of the American Horse Slaughter Prevention Act by the United States Congress would mitigate the need for individual state laws banning the slaughter of horses for human consumption. Never has that act been closer to being passed than today.

By the spring of 2004, more than 220 cosponsors in the U.S. House of Representatives had signed on to the act, which was reintroduced and sponsored by Representative John Sweeney of New York. Sweeney's congressional district includes Saratoga Springs, home of America's most prestigious and oldest Thoroughbred track, Saratoga Race Course, one of the nation's oldest harness tracks, Saratoga Gaming and Raceway, and the National Museum of (Thoroughbred) Racing and Hall of Fame. "Horse slaughter is inhumane, and the vast majority of Americans agree with what we're trying to do," said Representative Sweeney.

His bill, which was sent to the House Agriculture Committee and has been held up there for nearly two years, needs only 218 votes to pass the 435-member House.

On March 16, 2004, film star Bo Derek, an experienced horsewoman who keeps five Iberian warmbloods at her home in California, traveled to Washington, D.C., to speak to as many legislators as she could in Congress to help end the slaughter of horses in America. She had already made a short anti-slaughter video that begins with her, on horseback, talking about Monetary Justice's last-minute rescue.

"Americans don't raise horses for food and we don't eat them," Derek said in Washington. "Tens of thousands of horses are cruelly slaughtered in this country every year just so someone in Europe can have a fancy dinner. It's just plain wrong. I want to help end this horrid practice."

On April 27, 2004, Senator John Ensign of Nevada, a veterinarian and a Republican, and Senator Mary Landrieu, a Democrat from Louisiana, introduced the American Horse Slaughter Prevention Act's companion bill in the Senate, demonstrating bipartisan support for the legislation.

"A series of recent polls show that Americans overwhelmingly support a ban on the slaughter of horses for human consumption," Ensign said. "This should come as no surprise because horses in the United States are not raised for food. Despite state laws banning this barbaric practice, the slaughter of horses for human consumption continues because of an absence of a strong federal law to prohibit it. The time for such a law is now."

Landrieu noted, "Horses hold a special place in American history and culture, be it through farm work, racing, trail riding, companionship, or a host of other activities. Americans don't consume horsemeat, so most are shocked to find out this takes place at all. In fact, surveys from around the country showed that Americans overwhelmingly want this brutal practice to end."

In the House, Representative Ed Whitfield of Kentucky is doing all he can to speed up the process. "This is one of the most important things I'm working on," Whitfield said. "It's definitely the right thing to do. When I read about Ferdinand being slaughtered in Japan, I was shocked that a horse who had won the Kentucky Derby, who had won the Breeders' Cup, and been selected Horse of the Year could have ended up being slaughtered."

But, like many Americans, Whitfield was unaware that the slaughter of horses for human consumption continues in the United States. Said Whitfield:

> Then I came to realize that there are fifty thousand horses slaughtered per year in the United States. I didn't know that. Then I became interested in the transportation of the horses to slaughter. Those regulations specifically state that they recognize horses will be injured in the transportation process as it relates to double-decker trailers. The U.S. government knows injuries occur in transit. Then the [captive] bolt system used to kill them is being administered by untrained personnel. I've seen the slides of this and it's just inhumane the way it's being done.

The act was first introduced in the House by Maryland's Connie Morella (no longer in office) on February 14, 2002. At the time she said, "The American public is largely unaware that our horses are slaughtered for human consumption, and the three foreign-owned slaughterhouses operating in the U.S. would like to keep it that way. Americans do not eat horses. We do not raise them for food. The vast majority of Americans, when told that horses are being slaughtered for dinner in Europe, are horrified."

On February 13, 2003, Sweeney reintroduced H.R. 857, the American Horse Slaughter Prevention Act. In the following sixteen months, more than 220 of his colleagues signed on for the battle to end horse slaughter in the United States.

The bill has been held up since February 2003 in the House of Representatives Agriculture Committee, chaired by Robert Goodlatte, of Virginia. The bill cannot be voted on by the House of Representatives until it gets out of committee.

Sweeney calls Goodlatte his biggest obstacle in getting the bill to the floor of the House. When contacted in the spring of 2004 about Goodlatte's position on horse slaughter, one of his senior staffers charged with keeping the congressman abreast of the issue was surprised to learn that there was a slaughterhouse in Illinois trying to reopen. The staffer said of horse slaughter, "It's a complicated issue which involves emotion."

In his home state, Goodlatte is under pressure to act on the bill, but he has not budged from his position opposing the federal ban on horse slaughter. In a town hall meeting in Lexington, Virginia, on June 28, 2004, he attempted to defend his position in front of a room overflowing with his constituents at the Rockbridge County Administration Building. A few people brought their horses.

"He has basically stonewalled [the bill] and will not let it go anywhere," said Geri Wenz of Glasgow, one of the people who helped organize the groups that attended the meeting. "What makes him the authority on the issue? It appears Mr. Goodlatte prefers listening to another constituency, not his real constituents in his own Sixth District of Virginia. Texans may have helped you get your chairmanship, but they didn't elect you. We did. Stop the slaughter of our horses."

A survey conducted by McLaughlin & Associates of Alexandria, Virginia, and released two days after the meeting, showed that 74 percent of Virginian voters are opposed to slaughtering horses for human consumption.

Sweeney's bill would prevent the slaughter of horses in and from the United States for human consumption by prohibiting both the slaughter of horses for human consumption and the trade and transport of horseflesh and live horses intended for human consumption.

The highlights and rationale of the act are:

- Horses have played a significant role in the history and culture of the United States.
- Horses in the United States are not raised for food or fiber.
- As a nonfood and recreational animal, horses should be protected from slaughter.
- The foreign-owned horse slaughter industry has slaughtered and exported more than three million American horses for human consumption in the past two decades.
- Approximately fifty thousand American horses are slaughtered for human consumption annually in the U.S. by foreign-owned slaughterhouses. Tens of thousands of live horses are exported from the U.S. annually for slaughter.
- Horses slaughtered in these foreign-owned plants in the U.S. have often been hauled several thousand miles over several days, contrary to acceptable nonslaughter standards for water, food, and rest.
- Many horses shipped to slaughter are young, healthy animals. Others are old, sick, blind, crippled, and in otherwise poor condition and are unfit to withstand the rigors of long travel. Horses sent to be slaughtered are often shipped on crowded double-decker trucks designed for shorter-necked species such as pigs, cattle, and sheep, and are forced to travel in a bent position that can result in suffering, injury, and death.
- Killing of horses by foreign-owned slaughterhouses in the U.S. contrasts with the preferable method of killing by chemical euthanasia.
- Horses endure repeated blows to the head with stunning equipment that often does not render the animals unconscious. Some horses proceed, while still conscious, through the remaining stages of slaughter of being bled out and dismembered.
- Because horses in America are not food animals, veterinarians commonly prescribe and treat horses with potent drugs that may reside in horseflesh and be dangerous when consumed by humans.
- Because of the lack of disclosure on the part of the agents and dealers for the slaughter plants, people's horses are many times acquired and slaughtered fraudulently or through misrepresentation. Slaughter also provides a quick and evidence-free outlet for stolen horses.

- The imposition of a ban on the sale of horseflesh for human consumption, regardless of its source, is consistent with the international obligations of the U.S. because it applies equally to domestic and foreign producers and avoids any discrimination among foreign sources of competing products. Such a ban is also consistent with the provisions of international agreements, to which the U.S. is a party, that expressly allow for measures designed to protect the health and welfare of animals and to enjoin the use of deceptive trade practices in international or domestic commerce.

The act says a person shall not slaughter a horse for human consumption; import into or export from the U.S. horseflesh for human consumption or live horses intended for slaughter for human consumption; sell, barter, offer to sell or barter, purchase, possess, transport, deliver, or receive horseflesh for human consumption or live horses intended for slaughter for human consumption; or solicit, request, or otherwise knowingly cause any of those prohibited activities. The criminal penalty for violating the act, which could go into effect as soon as one year after its enactment, shall be a fine, one year in prison, or both.

The act was referred to the Agriculture Committee in February 2003 and, through the end of 2004, has remained there as its advocates try to bring the issue to the American public. They believe that the more Americans learn about horse slaughter in the United States, the sooner the act will be passed.

Bo Derek and actor Tony Curtis, whose wife Jill founded a horse rescue in Nevada, continued lobbying in Congress to increase support for the act and also have tried to raise public awareness of the issue.

Representative Whitfield is confident the end of horse slaughter is near. "I think there's a good chance," Whitfield said. "We have a number of obstacles, but we still have not made a full push for cosponsors. I feel we'll get 225 to 231 and it only takes 218 to pass. I believe it has support in the Senate, too. Our coalition is totally committed to passing it. It's definitely the right thing to do."

The Opposition

Frustrated by the act's lack of movement in the House of Representatives, Blue Horse Charities, formed by John Hettinger in conjunction with the Thoroughbred auction firm Fasig-Tipton Company—of which he is a major stockholder—took out a full page ad in the November 8, 2003, edition of the *Daily Racing Form* to coincide with a strongly worded letter sent to industry organizations, constituents, and the media. The letter says that the American Quarter Horse Association, American Association of Equine Practitioners (AAEP), and the American Veterinary Medical Association (AVMA) have cooperated to stall the act in the Agriculture Committee.

"When a bill is in the House and doesn't get out of committee and doesn't get hearings, someone has a thumb on it," the letter noted. "We would like to identify three groups who we feel, because of their stated positions, have their thumb on this one."

The ad included a cartoon by internationally acclaimed artist Pierre "Peb" Bellocq called "Fork in the Road." It shows a horse van heading up a road that branches off to a life of retirement and a road that leads to a slaughterhouse. The ad credited supporters of the act but chastised people "who profess to be anti-slaughter but do not back any legislative efforts and do not come up with any alternative solutions of their own."

So how could two groups charged with protecting animals, the American Association of Equine Practitioners and the American Veterinary Medical Association, not support the act that would eliminate the slaughter of horses in America for human consumption?

In "An Equine Veterinary View of H.R. 857 and Horse Slaughter in the United States," a paper circulated to congressmen in the winter of 2003–2004, the AAEP stated its position:

> Approximately 55,000 horses are slaughtered each year in the United States. These horses are most often sent to a processing facility because they are no longer serviceable, are infirm, dangerous, or their owners are no longer able to care for them. The AAEP, a professional association with a membership of 7,500 equine veterinarians and veterinary students worldwide, is dedicated to protecting the health and welfare of the horse and considers horse slaughter to be an important welfare issue for the entire equine industry.
>
> The AAEP has examined the transportation of horses to slaughter and their subsequent processing for over a decade. The AAEP's position on this issue states that the slaughter of unwanted horses at processing facilities is currently a necessary aspect of the equine industry in order to provide a humane alternative to allowing a horse to continue a life of discomfort or pain and possibly inadequate care or abandonment.
>
> The AAEP is not pro-slaughter, we are pro-welfare of the horse. Our association believes slaughter is not the most desirable option for addressing the problem of unwanted horses. However, if a horse

Illustration by Pierre "Peb" Bellocq.

> owner is not able or willing to provide humane care, the AAEP believes that euthanasia at a processing facility is a humane alternative to a life of suffering, inadequate care, and possibly abandonment.

The American Association of Equine Practitioners, however, said it would consider H.R. 857 if certain steps were taken, including federal funding of saved horses, whose care costs approximately $1,825 per year per horse. "This does not include veterinary and farrier care," the AAEP position paper states. "The AAEP is concerned that horse rescue and retirement groups in the United States will not have adequate resources without federal funding to meet a potential increase in owner 'give-ups'."

The second step the American Association of Equine Practitioners mandates for its "consideration" of support is the development of a specific enforcement plan to stop illegal transporters. Finally, citing an American Veterinary Medical Association recommendation made in

2000, the AAEP concludes that captive bolt slaughtering is considered "an acceptable method of euthanasia by veterinarians."

The AAEP position paper against H.R. 857 was circulated to members of Congress in the winter and spring of 2003-2004 as it became clear that the bill was gaining support. The AAEP advised its members: "The AAEP's Public Policy Committee has been working to educate legislators about the potential effects of this bill. Because the issue has such a tremendous welfare impact on the unwanted horse, it will also be beneficial for individual practitioners to directly contact their elected representatives. If you are concerned about H.R. 857's effect on the unwanted horse, the AAEP urges you to share your views with the senator and/or representative from your state or district. If you do not wish to draft a personal letter, the AAEP has provided sample letters that can be used for this purpose. The following address will take you to those letters, as well as the names and addresses of all members of Congress on the AAEP website: www.aaep.org/sample_letters.php."

In the April 2004 issue of *Equine Veterinary Education*, Larry R. Bramlage, D.V.M., the esteemed surgeon and AAEP president, did his best to settle the issue with his article "Banning the Processing of Horses for Human Consumption." He wrote:

> This is the most misunderstood legislation in recent memory. This bill is a poster child for unintended consequences. The argument for this bill is to improve the care of horses passing through public auctions and transported to processing facilities by banning the processing of horses for human consumption. The bill does nothing for the abuses and is harmful to horses as they pass through the public auction system and are transported. A recently passed [sic] transportation bill has a five-year phase in, of which we are in the third year. It has gone a long way to protect horses that are being transported to slaughter facilities. This bill negates this legislation.
>
> The bill does nothing for the unwanted horse that will still pass through the public auction at the owner's discretion except lower its value to near zero. It provides no money for the long-term care of the horses that are confiscated. Adoption and retirement facilities can currently be used rather than sending horses to auction, but they will be overwhelmed if all the horses that currently go to auction must be sent to adoption and retirement facilities. The crux of the bill is to ban the use of the carcass for human consumption. It says nothing about handling the use of the carcass for pet food. Consequently, the bill does

> nothing to improve the care of the live horse. In fact, because it reduces the value of the live horse, it tends to encourage abuse and lack of care.
>
> Curiously, the bill bans the use of captive bolt euthanasia, a method that has been endorsed by the AVMA and AAEP as humane and instantaneous. The bill allows the use of firearms, which the AVMA has termed only 'conditionally acceptable.' The uninitiated who use firearms to shoot the horse between the eyes simply destroy the horse's frontal and ethmoid sinus causing profuse hemorrhage, further suffering, and no lack of consciousness. It is odd that this bill should encourage this method as opposed to captive bolt euthanasia method, which requires operator familiarity with the equipment to make it work.
>
> In summary, there are many things wrong with this bill, but the curb appeal is so high that even college students are taking up the banner and marching for a cause they think they understand but have no real education about. The legislation on the surface seems good but is a horrible thing for the horse. In addition to the problems it causes for the horse, it is dividing the horse industry between breeds, a problem that we cannot afford if we are going to be an effective lobby in Washington. Veterinarians are one of the few groups of people who are able to shed some light on this problem for the horse owners.

The American Veterinary Medical Association, founded in 1863 and boasting a membership of more than 69,000 members, did its part to kill passage of the act when it issued the following press release on January 7, 2004:

Calling slaughtering "processing" does not make the act humane.Insisting that the deaths of horses at slaughterhouses is "humane" suggests that the people entrusted with the welfare of the horses have never bothered to watch the disturbing, yet readily available film of what happens repeatedly at horse slaughterhouses.

The American Association of Equine Practitioners and the American Veterinary Medical Association's assertion that the end of slaughter will bring rampant cases of neglect and abandonment is not supported by facts. According to the U.S. Department of Agriculture, in 1989, 342,877 horses were slaughtered in America. In 2003, there were 50,564, a difference of 292,313. The U.S. International Trade Commission says the number of horses exported to Canada and Mexico the past fifteen years has not changed dramatically.

Yet, in the last fifteen years, there have been just a handful of reported cases of neglect or abuse. The horse community effectively has been

Welfare Concerns Prompt AVMA Opposition to House Bill on Transportation and Processing of Horses for Slaughter

The American Veterinary Medical Association (AVMA) believes HR 857 . . . could potentially harm more horses than it saves. An analysis by the AVMA reveals that HR 857... fails to address the welfare of affected horses, fails to ensure levels of funding required to properly care for horses when humane slaughter is removed as an option, fails to recognize professional judgment in the appropriate application of methods of euthanasia for horses, and fails to consider potential environmental concerns associated with disposal of these horses.

"The welfare of animals, and, in this case horses, is the primary concern of veterinarians," said Jack O. Walther, D.V.M., AVMA President. "The proposed legislation, as written, could actually result in less humane treatment of these horses."

The AVMA outlined four areas of concern it believes must be addressed before it could consider lending support to the bill:

- Animal Welfare
- Funding for Humane Care
- Limits Placed on Euthanasia Methods for Horses
- Environmental Concerns Related to Horse Carcass Disposal

"Humane slaughter may not be the most desirable option for addressing the problem of unwanted horses," Dr. Walther said. "However, it may be preferable to allowing these horses to face a life of inadequate care or possible abandonment."

able to properly take care of hundreds of thousands of horses annually that are not sent to slaughter. Representative Sweeney shoots down the pro-slaughter forces' most frequent argument that there would be rampant cases of neglect and abuse if horse slaughter is outlawed. "Our opponents claim there is no organized process for adoption of horses," Sweeney said. "But there are networks of people all over the country dealing with horse adoption." In fact, there are now more than two hundred programs in forty different states and in Canada saving, retiring, and/or placing horses (see appendix).

Additionally, on November 5, 2004, the not-for-profit American Horse Defense Fund announced the publication of a 330-page resource book, which includes state-by-state listings of rescue organizations and equine sanctuaries. The book is available without charge, though a $10 contribution is suggested. The only fee is for shipping and handling. The AHDF's website is www.AHDF.org.

When a single Thoroughbred was found abandoned in rural Washington, it made national headlines. When he was saved from slaughter, that story also made its way around the country.

Why don't officials with the American Association of Equine Practitioners and the American Veterinary Medical Association ask those who run the retirement programs and rescue operations if they would be able to handle an increased volume of fifty thousand horses a year if slaughter in America ended today? They would get an emphatic, positive response from everyone in the country.

"The only people who would feel it would be the TRF and we're ready for it," said Thoroughbred Retirement Foundation Executive Director Diana Pikulski. "We want it. Oh, God, yes."

But the American Association of Equine Practitioners and the American Veterinary Medical Association do not speak for every veterinarian. In fact, one of them, Senator John Ensign, is sponsoring the Senate companion bill of H.R. 857 to end horse slaughter in America. Another is Nicholas H. Dodman, D.V.M., a professor in the Department of Clinical Sciences at Tufts University. In a letter to CANTER (the Michigan-based Communication Alliance to Network Thoroughbred Ex-Racehorses), Dodman wrote about captive bolt slaughter:

> Given horses' propensity for alarm and panic, their size and conformation, it is hard to imagine that this technique is humane. My bet is that many of them require a second shot or perhaps a third to end their now miserable lives. I've seen this happen in Scottish slaughterhouses where expert slaughter men have missed the mark. Why would the AAEP and the AVMA support such an industry?
>
> Why don't the AAEP and AVMA get together to organize appropriate euthanasia stations around the country, rather than allow horses to be subjected to this hideous experience? Why don't they work to develop adoption programs such as CANTER for racehorses that is doing such stalwart work?

This veterinarian does not support the captive bolt for horses. I wonder how many members of the AAEP or AVMA hold similar views and why the AAEP is attempting to turn their heads by sending them a form letter to send to their congressmen.

The AAEP's approach is not just pragmatic acceptance of what they might view as an unfortunate necessity. It is the active endorsement in support for a practice that is inhumane to animals, and I don't understand why.

He is not alone.

CHAPTER

10

The Industry Responds

"Two or three years ago, we were at a crossroads," said Diana Pikulski, executive director of the Thoroughbred Retirement Foundation. "The industry was either going to complete the stewardship of their athletes or we were going to struggle along with the other charities. We passed that crossroads. Racing has taken the right role."

John Hettinger has helped show the way. "Nobody has been more resourceful in looking over the care of horses than John," said Barry Schwartz, chairman and chief executive officer of the New York Racing Association from 2000 through 2004.

"The guy is honestly a saint, nothing short of miraculous," said Sue Finley, an editor and television producer and a member of the Thoroughbred Retirement Foundation board of directors. "He's moved it forward light years into the public consciousness in a way that no one but him could have done."

Hettinger, chairman of the Grayson-Jockey Club Research Foundation, a major stockholder and director of the Fasig-Tipton Company, a member of the Jockey Club, a trustee of the Jockey Club Foundation, and on the board of trustees of the New York Racing Association, could be enjoying a quiet retirement in his early seventies. He has been breeding and racing Thoroughbreds from his Akindale Farm in Pawling, New York, a converted dairy farm in Westchester County north of New York City, for three decades, and was the breeder of Wake Up Kiss, the 2003 New York-bred champion turf female. Now Hettinger devotes most of his time to ending the slaughter of horses in America.

It is a mission he undertook after a three-hour conversation with Diana Pikulski and Sue Finley at his house in Saratoga

Springs in August 1999. Until that point, Hettinger was like many other owners who would rather not think about horse slaughter. On that fateful day, Hettinger was moved to action. "We talked for a couple of hours," Hettinger said. "As we got into it, I had always thought that it was hopeless and a necessary evil. Now I see people doing positive things about it."

So he did, too.

"He's a big player in the game but was still willing to say, 'This is wrong,' " Pikulski said. "When there's something so wrong in your sport and livelihood, you don't ignore it. You fix it, especially if you have the wherewithal to do so. He turned people around. He's now completely devoted to doing what he can for horses and stopping the slaughter of horses. He feels it's the worst betrayal of the horse. You spend the horse's whole life teaching him to trust you, and then he works for you and you send him to the slaughterhouse."

That is a foreign concept to Hettinger. "I know a lot of owners, and I have never, myself, known an owner who would sell a horse to a killer," Hettinger said. "No friend of mine did. But, somehow, several thousands of horses wind up with killers every year. I think most owners don't have any idea that that's what they're doing. They don't know they end up this way. They just want to stop paying the [training] bills."

Hettinger has acted to stop the slaughter of Thoroughbreds. In the August 11, 2001, issue of *The Blood-Horse*, Hettinger penned the "Final Turn" guest column on the magazine's last page, in which he concludes that, at the end of a Thoroughbred's "useful life, he should be retired, adopted, or humanely euthanized if no better alternative can be found. Anything else makes a mockery of the words we have used for centuries to describe our game: 'The Sport of Kings.' "

A month later, his not-for-profit Blue Horse Charities was born. He convinced his fellow directors at the Fasig-Tipton Company to match any seller or buyer who donated one quarter of 1 percent of the final auction price of the horse they were buying or selling to a fund to save horses from slaughter. In its first year in 2001, 226 buyers and consignors combined to donate $90,000. Blue Horse Charities distributed the entire amount to a total of seventeen programs that adopted 522 horses. In 2003, there were more than one thousand buyer and consignor contributors, which allowed donations to reach $342,660 in Blue Horse's first three years.

Hettinger was also the moving force behind the unprecedented action of Fasig-Tipton and other major Thoroughbred sales companies to unilaterally raise the upset price of any Thoroughbred sold at one of their auctions to $1,000, effective January 1, 2001, to discourage killer buyers who could make a quick profit by buying the cheapest Thoroughbreds at those auctions and then selling them to slaughterhouses. The major U.S. auction companies taking this action were Fasig-Tipton Company, Keeneland Association, Barretts Equine Limited, the Ocala Breeders' Sales Company, the California Thoroughbred Breeders Association, and the Canadian Breeders Sales. In an open letter to the Thoroughbred industry, the sales companies said, "As representatives of companies that have sold horses for decades, it is our belief that the Thoroughbred does not belong in the human food chain."

New York Thoroughbred owner-breeder John Hettinger has been a driving force behind national legislation to stop horse slaughter.

New York Racing Association's Schwartz, an owner-breeder himself, applauded the move. "I praise the sales companies," Schwartz said. "By raising the upset price, killers can't make a profit. It used to be they could buy a horse for $300 or $400 and get fifty cents per pound at slaughter and get back $500 or $600. There's a break-even point where the killers can't make a profit."

In early April 2002, thanks to the efforts of Hettinger and other industry leaders, the National Thoroughbred Racing Association, the Jockey Club, and the Thoroughbred Owners and Breeders Association formally signed on to stop the slaughter of Thoroughbreds in America. In a joint statement that ran in Thoroughbred industry trade magazines, they said:

> In recent months, bills have been introduced in the House of Representatives that would outlaw the slaughter of horses for human consumption. Though not our intention to become involved in any political maneuvering, we felt it incumbent on us to make a statement concerning our beliefs on horse slaughter. We are definitely opposed to the slaughter of Thoroughbreds and urge all those involved in the Thoroughbred industry to support our ever-expanding rescue and adoption efforts and

to work together to find humane means of dealing with the problems presented by Thoroughbreds no longer suitable for racing or breeding.

The National Thoroughbred Racing Association had already started its own program, the NTRA Racehorse Adoption Referral Program, set up in the summer of 2000 by Hettinger. The program acts as a clearinghouse and network for the adoption of Thoroughbreds, many of whom are referred to the Thoroughbred Retirement Foundation and other programs.

Hettinger, typically, wants none of the credit for any of these major steps forward. "I don't really feel like I'm due recognition," Hettinger said. "I didn't do anything noble. I just wish I had started doing it sooner. When we make people aware of it, less people will do it. Just doing it is your own reward."

The people on the front lines of the problem are appreciative of the support they now have. CANTER founder and President Jo Anne Normile received a donation of $8,084 from Blue Horse Charities in May 2002, for her Michigan-based placement service. Normile wrote to Blue Horse Charities: "Your unexpected letter offering assistance for horses we adopted out in the year 2001 was the best present I could have received. It is not just the monetary assistance, which is desperately needed, but knowing that a national charity of the Thoroughbred industry had acknowledged our work is deeply appreciated by ourselves and myself. You have no idea how much that means to any of us in the trenches working with these horses."

Hettinger calls the American Association of Equine Practitioners' position that slaughter is a necessary aspect of the equine industry "absurd and heartless."

"The people that say this are forgetting the meaning of personal responsibility," Hettinger said, "and they are counting on other people doing their dirty work for them to get a few hundred dollars. That's all that this is about: money. The issue is five hundred dollars a head."

Breeders' Cup President and National Thoroughbred Racing Association Commissioner D. G. Van Clief Jr. agrees. "It's an issue of responsibility," Van Clief said. "Every owner of every horse should be willing to accept the responsibility throughout a horse's lifetime, including euthanasia if necessary. My main concern revolves around the conditions of the horses being transported and the lack of ability to ensure horses are slaughtered in a humane matter. These horses are at risk of being treated inhumanely."

But Van Clief, like Hettinger, sees the end of horse slaughter in sight. "Given the fact that horse slaughter occurs in three plants only, it's not a huge business," Van Clief said. "Therefore, the momentum ought to shift away from the slaughter option and toward personal responsibility. I think the Thoroughbred industry is eminently capable of doing the right thing."

Many in the industry already are. And they are not alone.

Others at Work

Prominent breeders in harness racing are also stepping up to end horse slaughter in their industry. Russell Williams, the vice president of Hanover Shoe Farms, president of the Standardbred Horse Sales Company, and vice chairman of the board of the United States Trotting Association, and Alan Leavitt, president and general manager of Walnut Hall Ltd., are leading the way. "I'm involved for the reason that any horse lover should be involved," Leavitt said. "Their lives shouldn't end tragically."

With Leavitt's help, the Hambletonian Society, which oversees the sport's greatest race, the Hambletonian for three-year-old trotters, unanimously passed a resolution on March 14, 2004, supporting the federal legislation to ban horse slaughter. Leavitt and his wife, Meg, got involved in the issue through their friendship with Nick and Kim Zito, and are also connected with U.S. Senator John Ensign of Nevada, who sponsored the anti-slaughter companion bill in the Senate.

"Nick got involved through my wife, who is still a Nevada resident and founded a private school in Las Vegas, the Meadows," Leavitt said. Leavitt deflects credit given to him in the anti-slaughter campaign by saying: "What John Hettinger has done is just unbelievable. Every good cause should have a John Hettinger. And Russell Williams should be canonized."

In 2002, Williams instituted a $1,500 upset price at the important Harrisburg sale in November, when some two thousand Standardbreds are sold annually. In two years, Williams bought thirty horses at bottom-rung prices and adopted out every one of them.

"In *Seabiscuit,* they said every horse is good for something," Williams said. "That's true. This puts pressure on the owner of the horse, which it should. The owner needs to think again, 'What do I do with this horse?' That's been the problem all along. There's not enough responsibility exerted. I'm hoping this will make Standardbred interests aware of the importance of getting the anti-slaughter legislation passed. As Tom Meeker [president of Churchill Downs Inc.] would say, 'We don't have to give another black eye to ourselves.' "

Williams was given an undercover film taken inside one of the Texas slaughterhouses. It sat, unviewed in his house, for months. "It took me about a year to watch it," Williams said. "It's heartbreaking. That's why you don't want to watch it. It was pretty bad. It showed that death is not instantaneous. It showed that the bolt gun doesn't always work properly. It showed that the people handling the gun haven't been able to find work anywhere else. There's one where the guy misses, and the horse turns toward him and looks at him and then the guy shoots him. That's after being trucked in. God knows what they've been through on the ride.

"It's not acceptable. That's why we have to do something. Nobody who has been there can say it's humane. No one can say it's okay. It's not about the horses. It's about us. The thing that kills me about the whole matter is I live on a breeding farm. This is my work. We teach these animals to trust us. What does it say about us as human beings if we don't remain true to that?"

CHAPTER 11

Making It Happen

In January 2000, a Thoroughbred named Make It Happen was about to become a gruesome statistic. After winning five of fifty starts and earning $40,228 in a five-year racing career, Make It Happen had seemingly been saved from slaughter by Michigan-based CANTER, the Communication Alliance to Network Thoroughbred Ex-Racehorses, which was started in 1997 by Jo Anne Normile and her husband, John.

CANTER had placed Make It Happen in a new home in a children's camp. But unknown to CANTER, the owners swapped Make It Happen for some construction work at the camp. Make The new owners then traded him for a pony headed to an auction in Kentucky where horses are often sold to slaughter.Make It Happen took the pony's spot on the trailer, and, at the auction, was sold for $390 to killer buyers from the Beltex slaughterhouse in Kaufman, Texas. His fate seemed sealed.

An anonymous e-mail to Jo Anne Normile tipped off Make It Happen's dire situation, but 11 days had passed. She contacted Shane Spiess, a Michigan Thoroughbred trainer who immediately called the trader who had sold Make It Happen for slaughter and told him that CANTER would pay $1,000 for the horse. That induced the trader to drive from Indiana to Kentucky to retrieve Make It Happen. When the trader arrived in Kentucky, Make It Happen was in a holding pen of doomed horses headed for slaughter in Texas that very afternoon. The man returned the horse to CANTER and was paid the bounty he had been promised.

But Make It Happen was not home free. While he was one of the exceptionally lucky ones to escape slaughter, his thirteen-day ordeal had left him two hundred pounds underweight. He had abscesses and fungus in all four feet. He had been bitten and gouged across his entire body and had a large, festering infected wound in his forehead above one eye. He was expressionless and lifeless and severely dehydrated.

Make It Happen was moved to the closest CANTER-approved foster home in Michigan to receive immediate veterinary treatment. After two weeks, he was strong enough to make the three-hour trip to Normile's farm in Salem Township, twenty miles northwest of Ann Arbor, Michigan. Six months later, Make It Happen had recovered. After passing a prepurchase veterinary exam, he was adopted by Mary Hejna of Chelsea, Michigan. Now they are teammates competing in United States Eventing Association competitions in Florida, Kentucky, throughout the Midwest, and in Canada. Make It Happen has a new barn name: Happy.

Make It Happen's experience is not unusual for horses headed to slaughter. It is the reality some people do not see and do not want to acknowledge.

The Normiles had their eyes opened soon after entering the sport in 1990. John, now retired, worked for the Michigan telephone company Ameritech. Jo Anne was a freelance court stenographer. Even though they had only been to a racetrack once (now-defunct Detroit Race Course), they owned a farm and decided to lease a Thoroughbred mare named Precocious Pat to produce two foals.

"Our business plan was to breed her twice and see if they were runners," Jo Anne Normile said. "And if they weren't, we would retire them to eventing."

They would never have sold the two foals. "We would never put a horse onto the ground and not take responsibility for that horse for the rest of their life," she said. "I take in any stray dog. If we take them into our house, we take responsibility. If you're the reason that a horse walks on the face of the Earth, you have a tremendous responsibility."

To her, it is a responsibility too many in the industry fail to acknowledge.

The Normiles' mare, Precocious Pat, produced a colt, Reel Surprise, a grandson of 1970 English Triple Crown winner Nijinsky II, and a filly named Scarlett Secretary, a granddaughter of American Triple Crown winner Secretariat. But the Normiles' experience as owners was brief and unpleasant. After twice switching trainers, Reel Surprise won a $5,000 maiden-claiming race at Detroit Race Course and then competed in low-level allowance races at that track.

"We lived thirteen minutes from the track," Jo Anne said. "As soon as our babies went to the track, I was there at six in the morning every day. I'd carry a condition book. These were our horses. We didn't tell people how to do their jobs, but if our horse got sick, he should go home. If there was any swelling, we'd have it x-rayed. We had a trainer, Pam Thibodeau, that knew that our horses were ours for all of our lives."

Make It Happen in his first competition in his new life as a jumper.

Being at the track every morning, Normile saw another side of racing. She said:

> When you're in the track kitchen or on the backstretch, you'd hear that 'someone' is coming to the track. There were two of them. Trainers approached them. What are you paying for them? They would say barn thirteen, stall A, and the horse was gone. I worried about that. What could I do? How could you make a difference? I was very frustrated. I thought about going to the press, but that wasn't the answer. That wasn't going to solve the problem.

Instead, she became involved in the Michigan Horsemen's Benevolent and Protective Association and was elected to the HBPA board for her first term in 1994. Two years later, on May 25, 1996, Reel Surprise broke down in a race and died on the track.

"There is nothing as exhilarating as watching a horse you bred and whose birth you videotaped come out of the starting gate; I don't care where they finish," she said. "But if you see your horse break down on the track, there's nothing more heart wrenching or devastating. Scarlett Secretary had raced three times without winning. After Reel Surprise died, we took her home and she never raced again."

Then the Normiles did exactly what they said they would: They converted her into an event horse. She would do well enough to be named Michigan Champion Novice Horse of the Year in 1997.

Their involvement with Thoroughbred racing could have ended when Reel Surprise died in 1996. But instead of walking away from the

Thoroughbred industry, Normile worked to change it. "I was very willing to donate my time," she said. "It gave me great access to the trainers to listen to them.

"Everyone knew we were no longer running our horses, but they wanted me to stay on the HBPA board. I wanted to stay on the board because I figured it would be easier to make changes if I was on the board."

One day in July 1997, she was standing at the rail when a trainer asked if she could come to his barn and look at one of his horses. "I immediately thought he wanted me to personally buy the horse," she said. "I felt an obligation as a board member to look at this horse and see what he had to tell me."

The horse was a tall, gray gelding that looked sound and had clean legs and no injuries. The trainer told her, "Look at this horse. He can't run a lick. But wouldn't the jump people you know be interested?"

Normile answered without hesitation. " 'Yes, I probably do know someone who might be interested.' That horse was sold immediately. The backside of the track is like a small town. The word spread that someone came two days later and paid well for a horse, much more than the going rate for horsemeat. I was bombarded. Walking from the kitchen to the rail, several trainers came up to me and said, 'Would you look at my horse?' A light went on: Why didn't I think of this? That was it. That's how CANTER started. I was on my merry way."

In 1999, as more trainers began to donate horses to CANTER, the group broke away from the Horsemen's Benevolent and Protective Association after it gained status as a not-for-profit 501(c)(3) charity.

Based at Great Lakes Downs in Muskegon, Michigan, CANTER began operation in September 1997, and it has grown exponentially with affiliates at tracks in Texas, Illinois, Ohio, West Virginia, and New England. New programs are being planned for tracks in Iowa, Indiana, Maryland, and Pennsylvania.

CANTER operates in two phases. In the first phase, volunteers at each participating track take photographs and record names of any horses eligible for adoption and put together an adoption list at CANTER's website, www.canterusa.org. Interested parties then contact the trainers directly. If a connection is made, arrangements are made to bring the adopted horse to his new home, one that must by approved by CANTER. Thoroughbreds coming off the racetrack are not for beginners and must be retrained.

"We've been criticized for not approving some of the horses' new homes," Jo Anne Normile said. "My response is, 'Where would they have wound up?' I think that's absurd. We're at a low-end track. Sometimes, these trainers need to sell their horses [to slaughter] to buy feed."

The second phase of CANTER's program is to rescue horses that have been on the list for a week or longer. CANTER takes in those horses even though it does not have its own farm. "We have people that are non-race people who'll take the horse in," Normile said.

CANTER helps to pay those horses' upkeep.

Unfortunately, there is no way she can include all horses or even all Thoroughbreds. "We're a breed-specific rescue [organization]," she said. "Our purpose is to be there for the racing industry when their horses need us." That means Thoroughbred ex-racehorses. The only exception CANTER makes is for track pony horses.

Normile quit her stenography career in September 2003 to work full-time with horses.

"It's seven days a week," she said. "There was no way I could do both."

While the program continues to expand—CANTER's 2003 budget for Michigan alone was more than $200,000—everyone involved in CANTER is a volunteer. "This is a non-paying job," Normile said. "I'm not paid. Nobody is paid."

Yet in one week in the fall of 2002, CANTER took thirty-two horses off the track at Great Lakes Downs, which is about a three-hour drive from the Normiles' home.

In the spring of 2004, eight saved horses from CANTER competed in an audition in Williamston, Michigan, to be part of the Netherlands Army's Queens Cavalry Escort of Honor for President George W. Bush, who was scheduled to be the grand marshal of Holland's seventy-fifth annual Tulip Time Festival celebrating the Dutch heritage prevalent in western Michigan. The horses included six-year-old Magi Magic, now owned by eighteen-year-old Julia McGuire of Williamston; Ronald and Clair Squires' Exalem; and thirteen-year-old Paddy's Landing, who earned $230,257 in his nine-year racing career while making 105 starts and is now owned by Heidi Rice of Mason, who adopted the horse for her son Stephen to ride.

Just sixteen days after the audition, CANTER received national exposure when the *New York Times* ran a front-page story about the Texas program (which has since been renamed LOPE Texas, LoneStar Outreach to Place Ex-racers). The article focused on Lynn Rorke, who operates the program on a twenty-six-acre ranch in Cedar Creek that she and her husband, Tom (a consultant to not-for-profit organizations), had purchased for $180,000. Donations cover only about $30,000 of their cost of operating the farm. Still, the Rorkes have placed ninety horses through its program.

In its first year, CANTER transitioned more than fifty horses to new homes and new lives. "Now we do over two hundred a year," Normile said. She estimates that CANTER has saved more than two thousand horses since its inception. Can more be saved? Normile has been astounded by the response from the Thoroughbred industry:

> I believe our racehorse placement program to be unique in the country in that it was founded by the racing industry itself. And I am very proud of the way they have embraced and used the program so enthusiastically. In the early years of CANTER, I could not even use the word 'rescue' in referring to the program for fear of upsetting those in the industry who didn't want anyone to think their horses needed 'rescuing.' Now they use the word! It's an exciting time to be involved in Thoroughbred rescue.

CANTER is unique in that it pays the equivalent of what the killers would for a horse that needs to be removed from the track if the owner wants to get rid of a horse he no longer can afford to keep.

"I will not walk past a horse in need standing in a shedrow simply because, financially, that poor horse's trainer will not donate him and must have meat price," Normile said. "We have been criticized for not making the trainers responsible and that we should refuse to purchase their horses for meat price. It is clear to me that those people demanding such action have not been in the shedrow of a track and looked into the eyes of a horse in that predicament."

CANTER relies on the Thoroughbred Retirement Foundation for help with funds. "We have a wonderful relationship with the TRF," Normile said. "We work well together because we are on the grounds of the tracks we service and are generally the first ones to know when a horse needs help."

Normile, like her compatriots who also save horses from slaughter, feels the end is in sight of horse slaughter in the United States:

> For the Chicken Littles who fear that banning slaughter will cause abandonment of horses on our freeways or prolonged suffering because horses will be neglected and left to starve, I offer that I believe Americans to be far more resourceful than that. I firmly believe in the adage, 'Necessity is the mother of invention.' In addition, we are a car-

> ing nation who has permitted this outrageous slaughter of our horses for the European food market only because the vast majority, even horse owners, were unaware of what really happens when a horse goes to slaughter. Once educated, they will provide some funding to help these horses, if prompted.

Some people do not need prompting. Normile relies on fourteen foster homes for the Michigan CANTER program and receives free surgeries from Michigan State University's Large Animal Hospital, which will remove bone chips and geld a horse when required.

Normile would like to see the American Association of Equine Practitioners and the American Veterinary Medical Association help the situation instead of condoning the status quo. "I would like to see the equine vets donate X number of euthanasias per year to a non-profit rescue agency," she said. "They are sure not to love that, but I gave up my job and donate my time and money, so why can't they, who earn their living off horses?"

Breeders also make their living off of horses. So do owners, trainers, jockeys, sales companies, racetracks, totalizator companies, and tens of thousands of other people in equine-related businesses. They can all contribute to the end of horse slaughter in America.

Just look at the example of one woman in Michigan who made it happen.

CHAPTER 12

Recycling

What is less humane for a champion horse, being slaughtered at a meat plant overseas or starving to death in a cattle field in America?

Our Mims, the 1977 champion three-year-old filly bred and owned by famed Calumet Farm, was never in danger of being sold to slaughter. She was, however, a poster horse for neglect before being rescued by her biggest fan, Jeanne Mirabito, and by ReRun, the not-for-profit Kentucky organization founded in 1998 by Shon Wylie and Lori Neagle.

ReRun is living proof that two people can have the same good idea at the same time. Wylie's husband, Jerry, a trainer who also worked on the starting gate at Turfway Park in Kentucky, was talking with jockey Jack Neagle in the track kitchen one morning. Jerry Wylie mentioned that his wife, Shon, who worked in communications for Breeders' Cup Ltd., was interested in starting an adoption program to spare retired Thoroughbreds from slaughter. Neagle said that his wife, Lori, who was on the board of directors of the Kentucky Humane Society for companion animals, had the same thought. The two women got together and created ReRun, a not-for-profit organization that helps ex-racehorses find a second career through careful adoption with a willing new owner.

"Both of us had the same idea," Shon Wylie said. "Lori was in the Humane Society. We did the same thing for horses initially. It started out really slow, a horse here, two horses there."

It is not slow any more. Originally begun under the umbrella of the Humane Society, ReRun was incorporated as a not-for-profit 501(c)(3) charitable organization in December 1998 and six years later had programs operating in eight states: Kentucky, New Jersey, Virginia, New York, Arkansas, Illinois, Maryland, and Washington. Wendy Moon, a former writer for *Thoroughbred Times* who served as executive director of the national Thoroughbred Charities of America, heads the ReRun chapter in Maryland.

In the winter of 1998–1999, Wylie found herself spending more and more time with ReRun, even when she was in her Breeders' Cup office. "I worked for the Breeders' Cup for a year and a half and ReRun was getting so big and so time-consuming," Shon Wylie said. "I was taking more calls for ReRun at the Breeders' Cup than calls for the Breeders' Cup. I said this is a tough choice. I liked my job. It had benefits and all that. But honestly, I had to make a choice, and it wasn't fair taking time doing ReRun while I was being paid to work for the Breeders' Cup."

So she left the Breeders' Cup in February 1999. Lori Neagle left two part-time jobs. "There was no money to pay a salary at ReRun," Wiley said. "Lori and I split a salary. We just did it because we had to do it. It was either grow it or keep it running halfway. We took a leap of faith, an absolute leap of faith. If God wanted this work done, he'd provide. Now we have other employees, including our adoption coordinator, Laurie Lane."

ReRun's mission is simple: recycle ex-racehorses. "We are a program that places great emphasis on making these adoptions work the first time," Wylie said. "Our niche is adoption and second careers."

Mirabito had been a fan of Our Mims a long time before she ever met her. "Our Mims, she was one of my heroes," Mirabito said. "She was just this great big, determined, beautiful girl. Seeing her in a news clip, I thought she was the most beautiful horse in the world. I would see pictures of her and every time it took my breath away."

She saw a different picture when she bumped into Our Mims at French Retreat Farm in Paris, Kentucky. A transplanted New Yorker from Hilton, a little town near Rochester, Mirabito was still getting used to a life in Kentucky she had never envisioned when she first met Our Mims. Mirabito, an elementary school teacher's assistant and writer who was working as a groom at the farm, recalled:

> I didn't know she was there. I was working on one side of the farm with four mares with their foals. The owner of the farm asked if I could help get the foals ready for the sales. When they moved the foals to the other end of the farm, I moved with them. One of the first days there, I met Our Mims in person. She kicked at me and hit the fence post and

broke it. And I said, 'Who's that?' They said, 'That's Our Mims. She's a witch.' And she was. She was aggressive and angry, so angry. She was hungry. She never forgot who she was. She was an Eclipse Award winner and here she was not being brought in for grain with the other horses.

Born on March 8, 1974, at Calumet Farm in Lexington, Kentucky, Our Mims had a very unusual racing career in that she never won a maiden race—which is restricted to horses that have not won a race—yet still became a champion. She was by Herbager out of Sweet Tooth, by On-and-On, making her a half sister to the brilliant colt Alydar, who became a legend for his rivalry with 1978 Triple Crown winner Affirmed. Both Our Mims and Alydar were bred and owned by Calumet Farm and trained by John Veitch.

Though she was winless in seven starts at age two, at three she blossomed into a champion. She won an allowance race at Hialeah Park in Florida, the Fantasy Stakes at Oaklawn Park in Hot Springs, Arkansas, the Coaching Club American Oaks at Belmont Park, the prestigious Alabama Stakes at Saratoga Race Course, and the Delaware Handicap at Delaware Park, on the way to being named champion three-year-old filly. She did not win those races on talent alone, but rather on determination. She won the Fantasy and the Alabama by just a neck and the Delaware Handicap by a nose. She finished her career with a record of six wins, six seconds, and a third in eighteen starts and earnings of $368,034.

Our Mims's progeny did not accomplish much on the racetrack, but their progeny certainly did. Our Mims's first foal, Heavenly Blue, by Raise a Native, was unraced but produced Play On and On, herself the dam of major stakes winner Continuously, who starred in England before returning to North America to win a major stakes race at Hollywood Park. The third foal produced by Our Mims was Mimbet, the dam of $1,100,567-earner Elmhurst, whose eight career victories included the one-million-dollar Breeders' Cup Sprint in 1997.

The success of her granddaughters and her own success on the racetrack did not seem to matter when Our Mims reached age twenty-one and could no longer get in foal. After being bred and not conceiving for five straight years, Our Mims was put out to pasture and left to starve in a cattle field.

Long since sold by Calumet Farm after its dispersal in 1992, the former champion wound up at French Retreat Farm in Paris, Kentucky, where one of her contemporaries, 1976 champion sprinter My Juliet, also resided. The two mares were not being treated equally, however, and Mirabito knew why.

Rescued champion Our Mims with Jeanne Mirabito.

" 'Juliet' was still producing in '98 and 'Mims' wasn't," Mirabito said. "She'd been barren for several years. There is an idea that when a horse is done working, she should be put out to pasture. Mims was an aged mare. She was accustomed to being fed twice a day and having her boo-boos taken care of. They literally put her out in the pasture with cattle. It's not too awfully bad, but they need more than cattle feed. She was not being brought in every day. She didn't have vet care or farrier care. Because she was old, she needed more than she was getting."

So Mirabito bought grain herself and took it to her in the field, on days Mirabito was working at the farm and on her off days, too. "She got used to me being around," Mirabito said. "She initiated a bond and let me start grooming her. It took about a month of feeding her grain for her not to be aggressive with me. The owner of the farm said I couldn't feed her inside because she was so aggressive and could hurt the other horses."

Mirabito left French Retreat Farm in early 1999 and saved Our Mims's life later that year. "The owner's two barns burned down," Mirabito said. "I asked him to donate a horse to ReRun, and he gave me Mims. She needed vet work way beyond my means. I called Shon [Wylie]. My daughter Cassidy and Shon's daughter Brittany went to school together, and I knew about

ReRun. I called ReRun and I said, 'I have a chance to save an old champion. Can you help?' They said yes immediately."

"Mims came to us and she was old then, twenty-five, and not in very good shape," Wylie said. "She needed worming. She needed her teeth done. And some TLC, and specialized food for elderly horses. We did all that. Mims knew who she was and what she had done, and she demanded that you treat her like a queen. She wouldn't go out with other horses. If you touched her without her permission, she would snap at you. Luckily, at twenty-five, she was a little slower. She started picking up weight and looking good, though she was really arthritic. Jeanne and her daughter had decorated her stall with all her accomplishments.

"Our Mims would have probably fallen through the cracks without ReRun," continued Wiley, a Cincinnati native, who once worked in publicity at River Downs and Turfway Park. "Our Mims stands to me as an example of what happens all the time in the Thoroughbred industry: What have you done for me lately? We've seen that played out time and time again. ReRun is very proud to have helped that horse."

The timing was crucial. "She would have died," Mirabito said. "There's no doubt in my mind. The day Shon came to see her there were dead cattle in the field. We had had a severe drought. She was just bones. She would not have lived through the winter. It would not have been considered unusual that a twenty-five-year-old mare had died."

Instead, Our Mims got a new life, spending about four months in rehab at Wylie's ReRun farm. After recovering, Our Mims was adopted on February 9, 2000, by Mirabito and her husband, a professor in molecular biology at the University of Kentucky. He does not share his wife's passion for horses, but he has helped her turn their forty-two-acre, run-down tobacco and cattle farm on the outskirts of Paris, Kentucky, into a safe haven for eleven horses. None was treated better than Our Mims. "We gave her four good years," Mirabito said. "She was treated like royalty here. She didn't try to hurt me ever. She tolerated a lot of people, and she showed affection to a few, but very few. She had a tight social circle."

Our Mims also had a hidden talent: painting.

Wylie freely admits that ReRun's brilliant fund-raising idea of Moneighs—paintings literally done by famous Thoroughbreds—was not original. "We kind of ripped the idea off from a friend who worked for a fund-raiser doing dog paintings," she said. "If horses get something on their nose, they want to wipe it off. I was talking to Jeanne and said, 'Here's a no-brain idea. Do you think we could convince horses to paint, use nontoxic paint, and not kill them?' "

Mirabito thought the idea was so well conceived that she volunteered Our Mims to do one of the first paintings. The Moneigh Our Mims created was auctioned for $1,000 at the 2002 ReRun Mane Event fund-raising dinner and earned her the award of "Artist of the Year." One by the now-deceased stallion Old Trieste fetched $5,000 in 2002.

Kelly Linn, who oversees the Virginia arm of ReRun with her husband, Paul, explained how she got one of her saved Thoroughbreds, Swell Ain't I, to paint.

> We got some paper, some nontoxic, water-based paint and mint oil with peppermint extract. She likes peppermints. We squirted some paint on cardboard and put mint oil on it. She dipped her nose in the paint wanting to find the mint. She smeared it all over. And we praised her. And she did it again. We only did it once with her. She wasn't very interested.

Other Thoroughbreds are interested in painting. The popular New York-bred gelding Funny Cide, winner of the 2003 Kentucky Derby and Preakness Stakes, was so enthralled with painting that he did five Moneighs in a row. Two-time Horse of the Year Cigar took Moneighs to a whole new level by actually grabbing the paintbrush in his mouth and painting. Most paintings are sold at auction, but many wind up on eBay. Moneighs have been done by such notable horses as Congaree, Free House, Gato Del Sol, Chief Seattle, Ten Most Wanted, Awesome Again, Monarchos, Groovy, Candy Ride, Bienamado, In Excess, Kona Gold, Cryptoclearance, Devil His Due, Mr. Greeley, and Serena's Song. Each Moneigh comes with a lock of hair from the artist's mane and a photo of the artist.

"It's a great way for owners and trainers to contribute to the ideals of ReRun, not with money, just their horses' time," said ReRun's Mary Simon, who oversees the paintings. "It's a win-win for everybody."

Simon takes it a step farther. She will travel to a farm and, in exchange for a donation to ReRun, help any horse create his own Moneigh.

It is obviously unconventional, but who cares? ReRun's origin was unconventional, too. ReRun has come a long way from its humble start. Since its inception, ReRun has placed more than three hundred horses. "A lot of them have gone on to make great riding horses or pets," Jack Neagle said. "They can have second careers if people give them a chance."

As with the Thoroughbred Retirement Foundation, horses are not the sole beneficiaries of ReRun. In October 2003, ReRun began working with Bethlehem Farm, a Christian-based program headquartered in Paris, Kentucky, that works closely with the Race Track Chaplaincy of America to help women struggling with problems in their lives by having them work with horses. ReRun has taken that program a step further, using women riders from Bethlehem Farm to help Thoroughbreds prepare for a second career as pleasure or show horses.

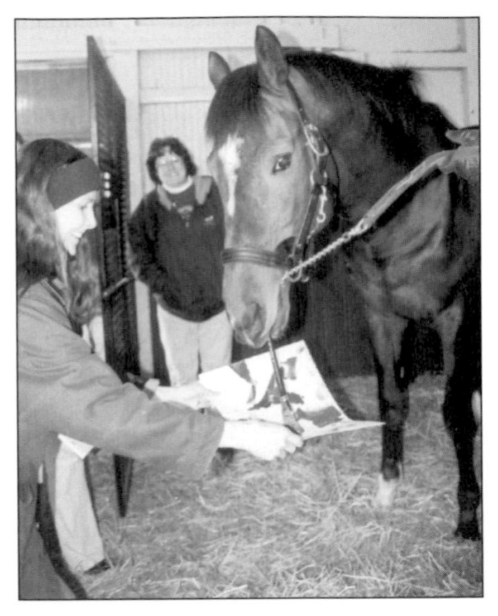

ReRun's Shon Wylie holds the canvas while Cigar paints a Moneigh.

Kelly Linn, met her husband, Paul, an engineer, through her work with ReRun, turned to horse adoption after a bad accident cut her riding career short. Said Linn:

> I started looking at horse rescues to see what was out there. ReRun impressed me just because of the background checks they did and what they did for the horses. I talked to Shon several times. It took about a year to find a horse I wanted to adopt.
>
> What impressed me was the length of the application I had to fill out and that they checked my references to make sure the horse was going to a good home and that I was able to handle an off-the-track Thoroughbred. They made sure I was financially stable. Was I trustworthy? Was I an honest person? They called the vet I was going to use. They asked a lot of questions of a lot of people to make sure I was the right person to take in a Thoroughbred. I got the horse, Regal Command, in January 2000. He's now ten. We adopted another horse last December, Find the Treasure, who's eleven.

The Linns, who have a 147-acre farm, started a ReRun program in Virginia in 1999 and have adopted out forty-seven horses. "We do everything, from updates on the website to contacting donors, to evaluating horses and what injuries they have."

Early in 2005, ReRun moved its headquarters to New Jersey, where Laurie Lane took over as director. In New Jersey, ReRun could take advantage of state funding and maximize finances by utilizing the program's Greener Pastures farm in Cecilton, Maryland.

Jack Neagle's wife, Lori, had a modest goal with her work in ReRun. "What I hope," Lori Neagle said, "is that there'll be less and less horses needing us as time goes by, if we can educate owners about doing the right thing with their horses, and reach out to as many people as we can."

She did not have to reach far to connect with Mirabito, who became a foster mom in the ReRun program although she lost Our Mims to colic on December 9, 2003. Our Mims was buried at Calumet Farm. Mirabito maintains a website, www.ourmims.com, to ensure the story of Our Mims will endure. The home page reads:

Famous Forsaken
Found Forever

"I came to Kentucky thinking 'I'm going to be in horse heaven,' " Mirabito said. "But some of the things I have seen are unbelievable. There are so many horses and so much neglect and so much mistreatment. Because there are more horses here, you see more bad things here than in other places. That's where ReRun, the Exceller Fund and, now, Old Friends are making a difference."

CHAPTER 13

Speaking Up for Horses

Susan Wagner's life with horses was born from watching Zorro on her TV set growing up in the Bronx. It was sealed some fifteen years later when she went to a racetrack for the first time and saw Secretariat win the 1973 Belmont Stakes by thirty-one lengths to become racing's first Triple Crown winner in twenty-five years. "I was majoring in theater in college, and I had never been to the racetrack before, but I wanted to see this horse," Wagner said. "He was my pinup. I went with a couple of friends and my mom to the Belmont Stakes. I stood against the front rail. I never budged the whole day. I was completely smitten. That horse was amazing to me. He was just the most gorgeous horse I'd ever seen."

She had seen other horses, having ridden since age five. The day after Secretariat's incredible performance, she showed up on the backstretch at Belmont Park to find a job walking hots. She has been caring for horses ever since, most recently through Equine Advocates Inc., the not-for-profit protection agency she founded in 1996, which has placed more than four hundred horses that might have ended up in a slaughterhouse.

"Nobody looks at a horse and says this is dinner," she said. "Every poll taken shows that people don't want to see American horses slaughtered."

In April 2004, Equine Advocates had sixty horses in its care waiting for adoption, twenty-four at Stonebridge Farm in Schuylerville, just outside Saratoga Springs, and the others dispersed at foster farms in New York, Ohio, Pennsylvania, and Canada. "We have a very strict contract with our horse adoptions," she said. "If something happens and the new owners can't take care of the horse they adopted, we take the horse back."

In late 2004, she opened a new facility, the 128 1/2-acre Safe Home Equine Rescue and Sanctuary in Chatham, New York, some sixty miles south of Saratoga Springs. "We've been doing this for nine years and this farm has been a dream of ours from the day I founded the organization," Wagner said. "We plan to turn it into one of the most effective facilities in the country."

Wagner has had a remarkable journey to reach this point in her life. She worked in New York as a groom for trainer Bobby Lake and spent nearly a year working at historic Sagamore Farms in Maryland. Then she began covering racing as a member of the media. Two of her freelance articles for the *Horsemen's Journal*, the magazine published by the National Horsemen's Benevolent and Protective Association, were cited for honorable mention for the Eclipse Award in feature writing—for a cover story on legendary jockey Angel Cordero Jr. in 1981 and a piece on Racing Hall of Fame jockey Braulio Baeza's attempt to become a trainer.

Wagner did radio and TV work for the New York Racing Association and the Teletrack off-track betting outlet in New Haven, Connecticut. In one story for the *Horsemen's Journal*, she journeyed to Panama to write about that country's legacy of great riders. A jockey she met there was about to come to New York and Wagner got talked into becoming a jockey's agent, the first licensed female agent in New York. Later, she became the agent for George Martens, who had ridden Summing to win the 1981 Belmont Stakes, denying Kentucky Derby and Preakness Stakes winner Pleasant Colony the Triple Crown. "I did it because I loved horses," she said. "I had a picture of every horse I ever won with. But to me, it wasn't rewarding."

When her mom became ill with cancer in 1987, Wagner left racing to nurse her for the next eight years.

But in 1989, after being out of work for a year and a half, Wagner needed a job. "I thought I would write or do public relations," she said. "I didn't want to go back into racing at that point. I just needed a job. I was having a hard time."

She could not find a job in public relations but finally found a position at the New York Zoological Society. "They said the only opening was wild animal keeper," she said. "I took it. Although it was very difficult for me, it was an experience I wouldn't trade."

That is because she met a horse there that brought clarity to her life. His name was Gandalf. "They brought this horse in," Wagner said. "Gandalf was part Thoroughbred and part Quarter Horse. He had been used at a riding academy. He was crabby and stubborn. Kids bothered him."

Gandalf became Wagner's first saved horse. She said:

> I had heard he would be sold for meat. I said, 'What do you mean?' I started researching it. I said, 'Oh, my God.' I spent all those years in racing and didn't realize it was happening right under my nose. I had heard about it, but it was such a secret. It was something that was done that was never talked about. It started to bother me. I started thinking, 'How can I do something?'

The first thing she did was to convince a former boyfriend who had a small farm in Maryland to take Gandalf. "He picked up Gandalf and another horse, a miniature horse in Staten Island named Rain Man," Wagner said. "He took both of them back to Maryland. I was trying to figure out what I wanted to do with my life. I couldn't quite figure it out. Then I learned about the slaughter issue."

She found her direction. In 1994, Wagner took a job in the New York office of Connecticut-based Friends of Animals. "They hired me because of my experience with horses," Wagner said. "Within three days, I realized this is not what I'm supposed to be doing. I should be working for an organization that only does horses. You could work 24/7 and not get everything done."

While she was working for Friends of Animals, Wagner went undercover and did several investigative stories about horse slaughter in America. One was for the syndicated TV newsmagazine *Hard Copy* in 1995. "I got death threats," she said. "When one of the pieces was shown on TV, they had to cover my face."

Hard Copy received some 10,000 phone calls after the piece ran, and another 50,000 viewers answered the poll question at the end of the piece: "Should horse slaughter be banned?" Out of 50,154 responses, 46,441 (92.6 percent) said yes.

Then Wagner left Friends of Animals to start Equine Advocates. "We were going to rescue horses and use the issues surrounding these rescues to raise public consciousness," she said.

With her background in the media, Wagner was well aware of how much impact a well-done story could have, particularly if it is told on television. In 1997, Wagner accompanied reporters and cameramen from Fox News to one of the major slaughter auctions in the East in Lancaster County, Pennsylvania, which resulted in a two-part series.

Equine Advocates founder Susan Wagner and rescued Hank the Tank.

That same year, Equine Advocates went to court to purchase, rescue, and place twenty-seven slaughter-bound camp horses from a bankrupt resort called the Ponderosa Fun Park in Catskill, New York.

"Media coverage of this story went on almost daily for over a month," Wagner said. Equine Advocates raised the money to buy the twenty-seven horses, then found homes for every one of them in New York, New Jersey, Maryland, Pennsylvania, and Canada. Equine Advocates won a special Equine Protection Award from the New York State Humane Association for its rescue work.

In 1998, Wagner accompanied an undercover camera crew from the TV show *Extra* to a Premarin farm in Manitoba, Canada, where pregnant mares were literally milked for their urine. The piece aired in February 1998, for which *Extra* received the 1998 Ark Trust Genesis Award.

A phone call to Equine Advocates in 1998 saved a Thoroughbred named Harry, a grandson of Exceller, the stallion who was slaughtered in Sweden in 1996. "We rescued Harry from Rockingham Park," Wagner said. "The owner and the trainer made the call and asked if I wanted him." Left unsaid was that if Equine Advocates did not want him, Harry would have been sold for slaughter. "We got two other horses with him,"

Wagner said. "When Harry got off the van, he was emaciated. He was lame. He was a basket case. Now he's our boy."

Harry was later adopted but returned to Equine Advocates in 2001 when his owner could no longer afford to care for him.

"All of our adoptions have three stages," Wagner said. "Approval of application, including personal references, as well as vet and blacksmith information, is first. Once that is approved, there is a site check. They need to have a lot of land for these animals and the land can't have barbed wire fencing. If that is approved, the people must sign a contract agreeing to never breed, sell, or use the horse for commercial purposes. And the foals also fall under that contract."

Equine Advocates made two double-digit rescues in 1999, saving and placing eleven abused and neglected American Saddlebred horses in New Jersey involved in a bitter bankruptcy case in February, then, three months later, saving thirteen horses headed for a Pennsylvania slaughter auction. One of those thirteen became the Equine Advocates poster horse.

Her name was Paula, a Standardbred and former Amish "buggy horse" no longer able to perform her duties. "She was crippled and emaciated when we rescued her," Wagner said.

On June 28, 1999, Paula was shipped from her new home in Pennsylvania to New York, where she was operated on for free by the late noted equine surgeon William O. Reed, D.V.M. Reed and his team performed joint surgery to remove a bone chip from Paula's right front foot. She recovered almost completely. "We have a few Amish horses and what the Amish do is a crime," Wagner said. "They run these Standardbreds into the ground."

Another success story is Banner, a Thoroughbred with great bloodlines. Born in 1985, Banner was originally named Bloodstock, being a son of Graustark and out of the Buckpasser mare Buck the Tide. Banner was two days away from slaughter after being used as an inexpensive stallion in Ohio.

"He had been used as a stallion because of his breeding," Wagner said. "They were going to sell him. So I bought him, maybe for five hundred dollars. He went directly to OSU [Ohio State University]. We had him gelded and vetted and he stayed with a friend of mine for a couple of months."

Then Banner was sent to his new home, ninety-acre Stone Bridge Farm in Schuylerville, New York, just down the road from Saratoga Race Course. That is where Jeffrey Tucker, an investment manager and Thoroughbred owner and breeder from New York City, took his own stand against horse slaughter. "I've seen enough to know they [the horses] are helpless and have no protection," Tucker said.

Tucker has owned Thoroughbreds in partnerships since the mid-1970s. His best horse was Empravatar, a son of 1975 Belmont Stakes winner Avatar whose promising career and life were ended prematurely by illness. Maybe that is one of the reasons Tucker got involved with Wagner's Equine Advocates.

"She's as dedicated to what she does as anyone I've come across," Tucker said. "She truly loves the animals and has the drive to do something which is very difficult: save them, get them adopted, and make sure they all have good homes. It was easy for me to get involved."

Tucker purchased his farm in January 2002, and immediately took in ten horses.

Stone Bridge Farm was also host of the annual Equine Advocates Awards Dinner and charity auction, where the Safe Home Equine Protection Award is presented. The first year the award was instituted, Herb and Ellen Moelis, who founded Thoroughbred Charities of America, were honored. (After the first year, Equine Advocates came up with a fitting trophy for the award winner. It is a sculpture of Banner done by Jan Woods, whose sculpture of former President Bill Clinton is on view in the Presidential Wing of the Smithsonian Museum in Washington, D.C.) In 2003, the award went to Gary Biszantz, the chairman of Cobra Golf and owner of Cobra Farms in California, who started Tranquility Farm, one of the largest and most successful equine retirement and adoption centers in North America. The 2004 award went to John Hettinger. He was presented with the award on August 5, 2004, at the Equine Advocates Awards Dinner, which featured a performance by Willie Nelson.

At the 2003 awards dinner and auction, a fiberglass horse painted by renowned equine artist Pierre Bellocq, better known as "Peb," was auctioned for $31,000 on a telephone bid from Marlena Brody, owner of Gallagher's Steakhouse in Manhattan. Peb, whose caricatures and drawings have graced the *Daily Racing Form* for fifty years, titled his horse, "L'Etalon Vert," The Green Stallion.

Peb, who was honored with a display of his work at the National Museum of Racing and Hall of Fame in Saratoga Springs, New York, in 2004, was happy to play a part in raising money to help end horse slaughter in the United States for human consumption in Europe and Asia. His French origin and heritage makes his support more meaningful as France is one of Europe's largest horsemeat markets and because the teenage Peb ate horsemeat while the Nazis occupied France during World War II.

"The only meat we could eat was horsemeat," Peb said. "That was our diet. When I came to America, I was convinced there was no more horsemeat consumption in France or in Europe because I never saw horse butchers."

Then Wagner asked Peb to donate "L'Etalon Vert" to benefit Equine Advocates. "When I started this for Equine Advocates, Susan Wagner was telling me it's a shame because horsemeat is sent to France and to Japan," Peb said. "I was surprised and entirely unconvinced. I told her, 'Susan, I haven't seen any horse butchers in France.' "

A little research convinced Peb that horse slaughter for human consumption, indeed, continues in his native France, in other countries in Europe, and in Asia. When he visited France in 2004, he did see horsemeat being sold at a small butcher shop, and he has taken an active role in trying to stop the practice.

"Of course, I am an advocate," he said. "Stopping the selling of horsemeat in Europe and Asia is beyond my control. But what they do in finding shelter for horses, I'm in favor of it. I do whatever I can here and in France."

When most people who love horses find out that the slaughter of horses in the United States for human consumption continues, they react quickly and meaningfully to stop it.

But until the ban of slaughter of horses for human consumption becomes federal law, horses must rely on deeply committed individuals such as Susan Wagner. She is speaking for animals that cannot speak. She is literally an equine advocate.

CHAPTER 14

The End of PMU?

Pregnant mares and young foals have not been spared from the horror of winding up in a slaughterhouse in the United States or Canada. Since 1942, the drug Premarin (derived from the urine of pregnant mares and referred to as PMU) has been prescribed by physicians who are treating the symptoms of menopause in women.

Premarin is conjugated estrogen extracted from the urine of pregnant mares. After their urine is sold to pharmaceutical companies and they are no longer in foal, the mares become expendable. Many are sold to slaughter. Some are allowed to have their foals, and then both mother and foal are sent to slaughter. Others are saved, but Susan Wagner's Equine Advocates estimates that as many as one million pregnant mares and foals may have been slaughtered in the past sixty-two years. The actual number is hard to tell because Premarin was principally a dirty secret for decades, the existence of hundreds of PMU farms in North Dakota and Canada little known. Synthetic conjugated estrogens have been developed in the past five years, and may eventually replace Premarin. Meanwhile, humane organizations reach out to rescue the at-risk mares and their foals before they are sent to slaughter.

Products known as Premarin, Prempro, Premphase, and other estrogen-hormone replacement drugs have been used by millions of women. "Women were being given drugs whether they needed it or not," Equine Advocates founder Susan Wagner said. "It's been going on since the 1940s in Canada and North Dakota, but it wasn't known about until twenty to twenty-five years ago."

That is because the story behind the manufacture of those drugs is a disturbing one. Mares are taken to PMU farms and repeatedly impregnated. They then are placed on "pee lines," where they are tied by their necks and strapped in by harnesses as pouches and tubes connected to their urethras collect urine. At many PMU farms, the mares are kept in narrow stalls that prevent them from moving freely or lying down comfortably.

"That, to me, is nothing short of torture," Dr. Ray Kellosalmi, a Canadian physician and surgeon and horse advocate, told Keith Morrison of NBC News in a January 18, 2004, story. "It's subtle torture, but it is still torture . . . Horses evolved through millennia in wide open spaces. They're not animals that should be subjected to factory farming conditions."

There are more blatant consequences. Due to the maltreatment of the mares, foals are frequently born impaired. Non-impregnated mares are sent to slaughter, as are many of the unwanted foals. "Life for PMU mares is brutal," Wagner said.

In the 1990s, there may have been as many as eight hundred PMU horse farms in Canada and North Dakota, and as many as eighty thousand mares impregnated annually.

By the late 1990s, the demand for PMU began to decrease. The development of a synthetic conjugated estrogen offered alternative treatment to physicians and their patients, and other synthetic drugs have followed.

In 2002, a study done by the Women's Health Initiative showed higher rates of heart disease, stroke, breast cancer, and blood clots among women who took the hormone replacement therapy drugs, which is another strong argument for the abolishment of PMU farms. Wyeth Laboratories, which manufactures Premarin, has maintained that PMU farms are extraordinarily well run and that the farms must comply with a standard code of care and undergo monthly inspections. According to a Wyeth spokesperson, the number of women using hormone replacement drugs has dropped from 15 million to 8.5 million, and the company established a trust fund of $3.7 million to care for the horses and provide transportation for them when the company began cutting back on the drugs' production.

In a report in 2003, Wyeth said it had sharply reduced its number of PMU producers, which resulted in a 50 percent reduction of PMU mares. At that time there were 409 PMU producers in Alberta, Saskatchewan, Manitoba, and North Dakota.

"Let's put that in perspective," said Dr. Kellosalmi, who stopped prescribing Premarin and Prempro for his patients. "Wyeth has been making billions from this industry."

On its website in early April 2004, Wyeth described itself as "a global leader in pharmaceuticals, consumer health-care products, and animal health-care products."

Underneath product information for Premarin and Prempro, which is accompanied by the slogan "Go Low With Prempro" and "Go Low With Premarin," visitors are told: "Using less hormones than standard doses, low dose therapies provide effective symptom relief for:

- "Hot flashes that can disrupt normal daily activities.
- "Night sweats that can interfere with a good night's sleep, leaving you tired and irritable.
- "Vaginal dryness that can make intimacy uncomfortable.
- "Premarin (conjugated estrogens) Vaginal Cream can also help relieve this symptom."

Wyeth also suggests that Prempro in a dose of 0.3mg/1.5 mg also effectively increases bone mineral density and bone mineral content and helps to protect against early menopausal bone loss.

But that is not all Premarin and Prempro do. Under a warning in boldface type, Wyeth tells consumers:

- "Estrogens increase the chances of getting cancer of the uterus. A woman should report any unusual vaginal bleeding right away while taking these products. Vaginal bleeding after menopause may be a warning sign of cancer of the uterus (womb). Her health care provider should check any unusual vaginal bleeding to find out the cause.
- "Do not use estrogens with or without progestins to prevent heart disease, heart attacks, strokes, or dementia.
- "Using estrogens with or without progestins may increase a woman's chances of getting heart attacks, strokes, breast cancer, and blood clots.
- "Premarin and Prempro should be used at the lowest effective dose and for the shortest duration consistent with a woman's treatment goals and risks. . . . If a woman does not have symptoms, non-estrogen treatments should be carefully considered before taking Premarin and Prempro solely for the prevention of postmenopausal osteoporosis."

That is the short version of risks. Wyeth's website has a thirty-one-page document detailing risks associated with these drugs.

Wyeth does not tell visitors to its website of the effects the two drugs have on the pregnant mares who have been producing the urine that has been used in their manufacture for more than sixty years, or what happens to the mares and their foals after they are no longer of use to Wyeth.

"The money from Wyeth is a drop in the bucket compared to the enormity of the problem and the number of horses that need to be rescued," Susan Wagner said. She continued:

> There are thousands of unwanted mares now. Slaughterhouses in Canada are overpacked. They are being sent to Beltex [the slaughterhouses in Texas]. You do this so long and you still can't believe it, it's so horrible. I still have horses in Canada that need homes. A lot of the rescue groups are involved. I think we have an obligation to do that because it's something that's winding down. I believe that in two or three years, there won't be a Premarin industry. There will be no need to use horses to make it when you can produce synthetic drugs. There are women who really suffer from menopause, but they don't need to take drugs from horses' urine. Thank God, this industry is winding down.

It may not happen for a few years, but until there are no longer any PMU farms, thousands of pregnant mares and foals need to be placed in new homes. Otherwise, they will be sold to slaughter.

That is why Equine Advocates rescued forty-six PMU mares from November 2003 to March 2004. Four of them arrived from Canada at Jeffrey Tucker's Stone Bridge Farm in Schuylerville, New York. At first, the horses would only eat hay. They had to learn to eat grain, which has essential nutrients and is basic feed for most horses.

Lisa Ribons of Saratoga Springs adopted Jill, a black Percheron cross, and Charlie, a rescued horse who was not at a PMU farm. Two gray PMU mares, Greta and Garbo, headed for a 3,000-acre ranch in New Mexico for retirement. Bernadette, a white Percheron, found a new home in New York. A shipment of thirty more horses from Canada arrived at Stone Bridge Farm in March 2004. On April 25, 2004, Jill foaled a filly named Lily. At 10:03 that very same night, Bernadette foaled a filly named Suzie. Two new lives came from two fillies that could have lost their lives in a slaughterhouse.

The sorry fate of PMU mares and their foals was a reality thirty-two-year-old Gina Brown in Brandon, Vermont, came upon by chance one day when she was on the Internet. "I found out about PMU online," Brown said. "I stumbled across it on a website. I said, 'That can't be true.' "

She desperately wanted it to be untrue. She had ridden and competed in barrel racing and other equine events most of her life and she had attended animal sales. "I'd been going to livestock sales with my uncle since I was little," she said. "He used to buy goats. You see a lot of gruesome stuff at sales."

Rescued PMU mare Bernadette with her foal Suzie at Stone Bridge Farm in New York.

Like many before her, it did not take Gina Brown long to learn the truth about PMU mares and horse slaughter. Said Brown:

> That's so barbaric. I was outraged, not only as a horse owner, but as a horse lover and animal lover. The horse is the most exploited animal in history. Our country is what it is today because of horses. We used horses for war. We used horses for transportation. We used horses to plow fields. The world would not have evolved if it weren't for horses. They deserve better.

What could she personally do, she asked herself. Gina Brown turned that question around: What couldn't she do? "At the time, we had three horses," she said. "I just told my husband, 'We're going to Canada with a horse trailer and we're going to save babies from slaughter.' He said, 'Yeah, that's great.' I don't think he thought I was that serious."

He didn't.

"I thought she was nuts, I really did," said Jeff Brown, Gina's ally and husband. "She said she was going to raise the money and I said uh-huh. And she did it. She got the money through donations and by adopting the foals ahead of time. I was kind of amazed."

Jeff Brown has an auto repair shop, Browns Auto Repair, in Rutland, Vermont, but he says he works harder at their farm than he does in his shop. "I'm the kind of person who, if I didn't have something to do, I'd go crazy," he said. "The shop is more like a vacation. There's more work to do at my home than there is at my garage. Pretty much, our life consists of saving animals. The majority are horses, but we also have dogs, cats, chickens, pigs, and ferrets."

In the spring and summer of 2000, Gina Brown went to work. "I started asking around, asking other people if they would be interested in getting a PMU foal," she said. "And we raised $12,000."

Then, in September, she and her husband hit the road, taking two trailers two thousand miles to the Winnipeg Livestock Sales in Canada. She had been advised by a contact in Canada, an activist in the PMU industry who had received death threats, that Winnipeg was the best sale to attend. The Browns slept in their trucks at rest stops along the way to save money. "It was kind of fun for the first night but got old very fast," Gina said.

Driving fourteen hours a day, they reached Grand Forks, North Dakota, some seventy-five miles south of Winnipeg, on the evening of Sunday, September 10.

She was not ready for what she encountered when she reached her destination on the day of the sale, Tuesday, September 12, 2000. "Nothing would have prepared me for what I saw," she said.

> There were over eight hundred horses. The majority were babies. This was a slaughter sale. I bawled my eyes out in the corner where no one could see me. Then I snapped out of it. I said, 'I'm here to do what I wanted to do.'
>
> There were hundreds and hundreds of baby horses everywhere I looked. They were so frantic and scared. Their cries were loud, and at one point I had to cover my ears. They had been abruptly weaned from their moms just hours before and a lot of them attempted to nurse off each other. I felt numb as I walked around the pens in disbelief. I had promised myself I was going to be tough and keep my emotions in check, at least until after the sale.

Gina had done her homework. Knowing that rescuers frequently get blackballed out of slaughter sales, she took a realistic approach. To play in the world of killer buyers and sellers, she had to play their game. "It's very much of a boys' club," she said. "It's almost like an underground industry. A lot of people don't know what goes on. A lot of owners don't know what goes on. For me to get out there and tell these horse dealers and meat buyers that they're awful people wouldn't do the horses any good. My point of focus is the horses."

Jeff Brown was proud of how she handled herself. "I was nervous when she went to the meat auction, but she can take care of herself," Jeff said. "You don't go up to them and tell them they're the scum of the Earth and then buy horses from them. She went right to the meat auctions and bought them. I thought that was kind of cool."

It was not an easy assignment, especially for a novice. "It was tough," she said. "The sale was set up around the meat dealers. They have the tables. They have the phones. I had a list of twenty-four foals. We ended up buying thirty-two PMU foals and one pony that someone threw in. He was in pretty tough shape. We trucked them back here."

On the way home, the Browns were helped in Buffalo, New York, by Western New York Equine Sanctuary. "The folks there were absolutely fabulous," Gina said. "They arranged for our foals to be housed in an indoor arena. They provided our babies with good hay and fresh water. They also provided us with a place to sleep and treated us to dinner."

Gina had been worried that one of the foals, a draft colt named Eeyore, would not make it to Vermont. "Eeyore was malnourished from the start," Gina said. "When he walked, he looked almost animated, like a puppet. His huge hooves, long legs, and large head looked so out of proportion with the rest of his body. Eeyore quickly lost more weight and became weaker and weaker with every mile."

But Eeyore could wobble around enough to find his hay and water and survived. "As long as he was eating and drinking, I knew he would make it," Gina said. "He wanted to live. He was fighting to live."

And he did. All the horses made it to Vermont, and all but one foal found new homes. "We kept one," Gina said. "We keep one or two every year to do natural horse training. From the thirty-two, we placed thirty-one of them. It went wonderfully. We had a lot of response. The beauty of it is they're inexpensive. You can get the horse you always wanted and you're saving the life of a horse. The icing on the cake is you get that pretty Palomino you always wanted. I think, 'What can I do for the horse?' instead of, 'What can the horse do for me?' "

Eeyore survived, and, after "tons of vitamin and mineral supplements, good nutritional grain, three dewormings, and plenty of TLC," is thriving.

The next year, the Browns bought seventy-five foals from livestock auctions. In 2002, it was ninety-five. In 2003, they bought ninety-two PMU foals and placed ninety. "It doesn't even seem like a drop in the bucket," Gina said. "To see those ninety-two babies come home. They traveled five days. They've been torn from their mothers. And when they get here, it's the neatest thing in the world. I think they know. They seem at peace. Every foal we save is a piece of education. That foal will educate people all his life. People will say, 'He was a PMU foal.' And someone will ask, 'What's a PMU foal?' It's a real tribute in itself, saving just one foal."

Said Jeff Brown, "To see the foals afterward, it seems like they have a sense of appreciation. It makes you feel good that you were able to save that animal from abuse, neglect, or slaughter."

In doing that, there are always thoughts of despair for the horses that were not saved. When his wife feels down, Jeff reminds her of the fable they received from a friend and posted on a wall in their house. The tale is about a woman saving starfish on the beach by throwing them back in the sea after they had been stranded on shore. A man sees the woman and asks, "What are you doing?" She says, "The tide is low and they'll die if I don't throw them back in the sea." The man says, "There's a million of them. You can't possibly make a difference." She throws one in the ocean and says, "It made a difference for that one."

CHAPTER 15

Horses Helping Horses

Growing up in Brooklyn in the 1930s, how could Herb Moelis have imagined his legacy would involve horses? "Probably the only horses I ever saw were police horses," Moelis said. "I was always a sports fan, but I never really got into racing."

He is now. And he is leading the way in horse rescue and charitable work, showing how a few people can make a big difference. He and his wife, Ellen—a New York City girl from Queens—joined forces with Allaire duPont, their Delaware neighbor who campaigned five-time Horse of the Year Kelso, to make an impact on Thoroughbred racing that has been nothing short of remarkable. In 2003 alone, their Thoroughbred Charities of America raised and donated $1.5 million to nearly one hundred organizations. "Most of them are for horse rescue and retirement, but we also help backstretch people," Herb said. "We've expanded into that."

Thoroughbred Charities of America has been the common denominator in rescuing Thoroughbreds, providing children with a day care center at Belmont Park in New York, helping backstretch workers with drug problems in Maryland, enabling handicapped children in Kentucky to ride horses, helping students at the Oklahoma State Veterinary School learn about horses, and allowing troubled youth in California to see another side of life. All have been helped by this grandfather of ten, his wife, and the legendary duPont. Ellen Moelis says duPont is an inspiration.

"She's an incredible horsewoman, first of all," Ellen said, "and she's so accomplished in other things, flying planes, golf.

She's an all-around sportswoman. In addition, she has a very kind heart and she's a very generous person."

So are Herb and Ellen Moelis.

"It's a very good feeling," Herb said. "Probably the only negative was worrying about what happens when we're gone. What happens when this thing doesn't run? Ninety-five percent of the work is done by Ellen and myself."

The Moelises have already found a solution. In early 2004, Thoroughbred Charities of America formed a strategic alliance with the Thoroughbred Owners and Breeders Association based in Lexington, Kentucky. "That ensures continuity," Herb said.

That is important because he thinks America is on the threshold of ending the slaughter of horses for human consumption. "I really don't think we have a long way to go," he said. "I think we're very close."

He salutes the people leading the fight. "John Hettinger has done a marvelous job," Herb said. "I see it on my own board. We were anti-slaughter. We presented it to the board several years ago, and I was surprised to learn that many in the industry were not anti-slaughter. They had good reasons. They said there would be horses in the street. But now they've turned around."

Herb did not get involved in Thoroughbred racing until the early 1980s. After studying at the Wharton School at the University of Pennsylvania and New York University, Moelis practiced law and was a certified public accountant, eventually heading an equity leasing company. When he decided to sell National Guardsman, a division of the equity leasing company, to Irish concerns, he was introduced to Thoroughbred racing.

"There were attorneys from Ireland [representing the buyers of National Guardsman] who happened to be involved with horses," Ellen said. "They said, 'It's fun, it's exciting, and you can really enjoy it. It will be a new phase in your life.' "

Herb had always planned on retiring to a farm, so he was intrigued by the sport. "In my college years, I always wanted a farm without knowing what I wanted to do on a farm," he said. After buying his first Thoroughbred, he knew. "We bought a filly, La Reine Rose, in 1984. She won four of her first few races. It seemed like a real easy game." He laughed because he learned how difficult a game it is. It didn't deter him. "It gave me an idea: This is what I wanted to do on a farm."

He and his wife—the two met at Hemlock Farms in Pennsylvania in 1981—purchased a 160-acre property in rural Middletown, Delaware, in

1986, and named it CandyLand in honor of their grandchildren. "We had decided, since we were living in Manhattan, let's make a line and go south of New York," said Ellen. "It was going to be a weekend retreat."

Three years later, she sold her company, Allison Management Corporation, which managed apartments and co-ops. "I was president and sole owner," she said. "I'd spent thirty-eight years there. I sold it in 1989 basically so I could spend full time at the farm. I absolutely do not regret making that choice. Once in a while, I get a twinge for business, but I do all the books [for CandyLand] and that keeps me busy."

That is because the Moelises operate a stable of about fifty horses, including more than twenty broodmares, a dozen horses in training, foals, and three or four horses they retired, including La Reine Rose and Cat Be Nimble.

La Reine Rose and Cat Be Nimble are being cared for in their retirement, never in danger of an unhappy final chapter in their lives. The Moelises, though, quickly learned that was not a luxury for all Thoroughbreds. "We were rather new to the business, my wife and I, and we were not aware of it," Herb said. "We never had a thought of, 'What happens to other horses?' "

Ellen added:

> When we first got into the business, the horses you get to know are the big names. We went to Claiborne to see Secretariat. That was the first important stallion that I saw. We went to a lot of other farms. You kind of think all horses end up that way if you don't know the business. Then I started finding out about the other horses and the end they may come to.

Herb said, "Most of those other horses were going to a slaughterhouse or leading a terrible life after racing."

What pushed them into action was a magazine article written by their Delaware neighbor, Allaire duPont. "She had written an article about the terrible end to our beautiful horses' lives, not just horse slaughter, but that they also were maltreated," Ellen said. "They were starving. Their ribs were showing. And they were taken to New Holland [the Pennsylvania auction] and were purchased by the slaughterhouses.

"And I looked at my husband, and he looked at me, and I said, 'This is terrible.' I cry easy. I was crying when I read that article. We agreed that something should be done. We didn't know what could be done."

The Moelises decided to join the battle to rescue and retire Thoroughbreds to prevent them from being sent to slaughter. They joined the Thoroughbred Retirement Foundation, but they did not stop there.

"We were trying to think of an idea to raise funds," Herb said. "We were good friends with Allaire. Her farm is about two miles away from our farm. We decided we could raise money by selling stallion seasons [one-time breeding rights to a specific stallion] at auction. It would be horses helping horses."

One of the first women elected to the Jockey Club, the breed registrar for Thoroughbred racing, Allaire duPont was happy to help, having spent her life around horses. She had bred and raced Kelso, one of the greatest Thoroughbreds to ever step foot on a racetrack. Trained by Carl Hanford, Kelso was named after duPont's friend Kelso Everett. Kelso, the horse, was a muscular dark bay who started sixty-three times in his magnificent career and won thirty-nine races. He earned $1,977,896, a record that stood for fifteen years. Thirty of his victories were in stakes and thirteen of those were accomplished while carrying high weights of 130 pounds or more, conceding significant weight time and again. He set seven track records, none more amazing than the 2:23 4/5 clocking he set for the one and a half-mile Washington, D.C., International Stakes on November 11, 1964, at Laurel Park. Nine years later, Secretariat would complete his Triple Crown by taking the one and a half-mile Belmont Stakes by thirty-one lengths in a world record of 2:24 on dirt. Kelso had run faster. Kelso was Horse of the Year from 1960 through 1964, a period of domination never seen before or after. Joe Hirsch, the brilliant columnist for the *Daily Racing Form*, wrote, "Once upon a time there was a horse named Kelso. But only once."

Kelso retired to a life of riding and hunting on duPont's 900-acre Woodstock Farm on the banks of the Bohemia River in Chesapeake City, Maryland. DuPont raced in the name of Bohemia Stable. Kelso died of colic at the age of twenty-six.

But duPont loved all horses. In 1989, the Moelises and duPont held their first dinner-and-auction fund-raiser at the Moelises' farm to benefit the Thoroughbred Retirement Foundation. The eight or nine stallion shares on offer that night were, "Mainly from Maryland because we had not yet made inroads to Kentucky," Ellen said, and included stallion seasons to Two Punch, Allen's Prospect, and Waquoit. "Some people came to the party strictly to get a season to Two Punch because they couldn't get one commercially."

They also auctioned racing memorabilia, art, and an antique chair with needlepoint done by duPont.

Herb Moelis was delighted with the first auction. "The first year, we had about fifty people in our house and we raised about $15,000," he said. "My wife and a few friends did all of the cooking. It was very successful for a first-year venture."

With each succeeding year, the auction has been more successful. "When we reached $500,000, we thought this was an awful lot of money for one organization when other small organizations are struggling," Herb said.

Herb and Ellen Moelis, founders of Thoroughbred Charities of America, have raised millions to help horses.

After conferring with the Thoroughbred Retirement Foundation board, the Moelises decided to create Thoroughbred Charities of America as a separate entity while still giving some of the funds it raised to the foundation. They also moved the date of the auction from February to early December to make it more attractive to breeders to buy stallion seasons well before the traditional mid-February start of the breeding season.

In 2002, the Moelises decided to expand the annual dinner-auction, which had come to attract four hundred guests annually at their Delaware farm. The Moelises arranged for an audio simulcast of the dinner-auction to the Cross Gate Gallery owned by Greg Ladd in Lexington. So, in addition to the 380 people at CandyLand, 125 people attended the auction in Kentucky. "We decided to bring Lexington into it because a lot of our seasons are there," Herb said. "Greg Ladd let us use his gallery."

Stallion seasons usually sell for $500 to $5,000 at the auction. In 2002, the top-priced season went for $95,000 for the right to breed to the successful sire Unbridled's Song. A season to Unbridled's Song fetched the same amount in 2003, when 300 seasons and 250 pieces of equine-themed artwork were sold to raise $1.5 million.

Ellen Moelis had thought the 2003 auction was going to be a disaster. "It had snowed for three days," she said. "I remember getting up at 4:30 in the morning on the day of the auction to see the weather, and I said, 'It's snowing again. Nobody will make it here.' But we still got 380 people, and they came from all over, New York, New Jersey, Pennsylvania, Virginia, and Delaware. Some of the attendees banded together to rent a bus so they wouldn't have to drive."

It was a testament to the efforts of the Moelises and duPont that all those people braved the elements to come, but also to the depth of feeling those people have for horses. Said Ellen:

> I think, basically, that you become involved with charity for two or three reasons. You want to help when there is a need to help and when there is an injustice. My husband and I devote six months of every year to get this auction together and give out the awards. You don't stick with it if you don't get any satisfaction from doing it, and I absolutely do. I'm very proud.

At the Thoroughbred Owners and Breeders Association's annual awards dinner in Lexington, Kentucky, on September 10, 2004, Herb and Ellen Moelis were honored with that organization's Industry Service Award in recognition of their work to promote the well-being of retired racehorses and the people who work with them.

Not bad for a guy from Brooklyn and a girl from Queens.

CHAPTER

16

Fans Helping Horses

In 1982, Michael Blowen, a reporter for the *Boston Globe* and freelance writer, pitched a story idea to *Spur*, a Thoroughbred racing and breeding magazine, about Massachusetts's version of the Triple Crown: a featured Saturday race at each of the state fairs, Northampton, Great Barrington, and Marshfield. Blowen decided to follow Carlos Figueroa, a trainer based at Suffolk Downs in Boston, as he attempted to win each of the races with three different horses.

But writing that story is not what Blowen remembers most about the Massachusetts fairs:

> I was at the fairs when I saw these killer trucks pull up to take these horses to slaughter. It made me nauseous. You hear the horses make this unearthly sound when they get on. It's a sound you've never heard before or will ever hear again. Sometimes, they put horses on these trucks with broken legs.

That is not his worst memory. "I saw horses put in dumpsters," he said.

Blowen's story about the Massachusetts Fair Triple Crown ran in *Spur*, but Blowen did not write many other racing stories and did not run into Figueroa again for several years. Then, on a typically dreary, cold, and nasty February afternoon at Suffolk Downs, Blowen was at the track and bet on a simulcast race in Florida. He liked an outsider who won the race but was disqualified. "I lost $20 and would have won $2,000," he said.

Understandably upset, he took a walk down the length of the near-empty apron and ran into two people: Figueroa and outrider Kristin Prado, the sister of one of the nation's best jockeys, Edgar Prado. An outrider corrals a horse if he breaks loose from the post parade or on the way to the starting gate. "She looked like she was frozen half to death," Blowen said. "Yet I said to Carlos out of the blue, 'I'm sick of betting on horses. Can I come to work for you?' "

Figueroa said yes, but Blowen had to show up Monday at 6 a.m. He did. When he arrived, apparently not the first backstretch neophyte Figueroa had invited, a groom said, "Oh no, not another one."

Blowen, who had ridden horses a bit at age nine, was terrified of horses. But he worked briefly for Figueroa as a groom. One of the horses he took care of was Saratoga Character, a son of Saratoga Six who typified Figueroa's horses, a once-classy animal who had been injured and moved from the major circuit of New York to the minor leagues of New England.

Kim Zito, trainer Nick Zito's wife, saw Saratoga Character in a $4,000 claiming race at Rockingham Park in New Hampshire, and retired him to the Thoroughbred Retirement Foundation, which placed him at the TRF-affiliated Blackburn Correctional Complex in Kentucky, to work with prisoners. Blowen went to Kentucky to do a story on Saratoga Character for the *Boston Globe.*

Several years later, Blowen decided to take a buyout from the *Globe,* where his wife, Diane White, voted best columnist by *Boston* magazine, also worked. "The buyout came and I told my wife the first gift I was going to give myself was a trip down to see the 2001 Belmont Stakes," Blowen said.

At Belmont Park before the race, Blowen met Elinor Penna, the wife of trainer Angel Penna and one of the first female reporters to cover the New York Giants. Elinor Penna was on the board of the Thoroughbred Retirement Foundation and invited Blowen to a dinner held in conjunction with an auction the TRF was holding the night before the Belmont Stakes. At the dinner, someone asked the newly independent, then-fifty-four-year-old Blowen, "Why don't you come to work for the TRF as operations director?"

Blowen did not think the offer was serious. But it was, and when it was repeated the next afternoon, Blowen took it.

Originally, he worked out of his home in Arlington, Massachusetts, but he and his wife moved to Midway, Kentucky, to be closer to the Thoroughbred racing and breeding communities.

In Kentucky, Blowen met Rick Trontz, who owns and operates Hopewell Farm near Lexington, the home of 1998 Horse of the Year

Skip Away and 2002 Breeders' Cup Classic winner Volponi. Trontz mentioned the high daily cost for boarding stallions after their breeding careers are over. That gave Blowen an idea.

"We don't promote these horses enough," Blowen said. "When you go over to the [Kentucky] Horse Park and see [two-time Horse of the Year] John Henry, it gives you a chill. I thought you could have a place for retired stallions. The idea was to do it just like the Horse Park so that no one would forget how great these athletes were."

Michael Blowen formed Old Friends, a not-for-profit organization to do exactly that, in June of 2003. "Two weeks later, people learned that Ferdinand died in a slaughterhouse," Blowen said. "Then everything exploded."

Blowen originally considered locating his stallion retirement home near Saratoga Race Course in upstate New York. Both Dennis Brida, president of New York Thoroughbred Breeders, and Saratoga Springs resident Jack Knowlton, the managing partner of Sackatoga Stable, which campaigned 2003 Kentucky Derby and Preakness Stakes winner Funny Cide, thought it was a good idea. But the plan changed when Blowen stopped at an antique shop in Midway owned by Betty Sue Walters. "I kind of got frustrated with the way things were going in Kentucky," Blowen said. "I moaned to her that all our contributions were coming from everywhere except Kentucky. She said, 'Meet me tomorrow.' "

They met the next day at Afton Farm, owned by Betty Sue and Philip Walters. "They said, 'What would you think of having it here?' " Blowen said, and he took them up on the offer. The Walters lease land to Old Friends on their Afton Farm for $1 a year. The first horse saved by Old Friends was not a stallion, but a twenty-one-year-old daughter of Exceller.

Thanks to the work of Howard Keck Jr., the son of Ferdinand's breeder, and U.S. Congressman Ed Whitfield and his wife, Connie, people are able to track mares who had been sired by Ferdinand, Exceller, and other Grade 1 stakes winners. That is how Blowen learned that a filly by Exceller, aptly named Narrow Escape, had not received a single bid at a dispersal sale of John Franks's Estate, which had been conducted by the Fasig-Tipton sales company. Blowen was ready to pay $1,000 to buy the mare, but Fasig-Tipton donated her in February 2004.

"She's got a great personality," Blowen said. "We were interested in getting a horse who would symbolize what we're trying to do. Our goal was to make this worthless horse priceless."

TV stations and newspapers loved the story and the publicity from the rescue generated a lot of mail. In order to raise funds for Old Friends, Blowen decided that he would issue certificates signed by great

Narrow Escape, a daughter of Exceller, runs in the back of the pack after being donated to the Old Friends program.

horses' trainers to anyone who donated $50 or more. Blowen sent out and received back signed certificates from Nick Zito, trainer of Kentucky Derby winners Strike the Gold and Go for Gin, and Mack Miller, who trained Kentucky Derby and Travers Stakes winner Sea Hero. Both Strike the Gold and Sea Hero stand at stud overseas and will be sent to Old Friends when their breeding careers are finished, ensuring that they do not suffer the same fate that Ferdinand and Exceller did.

Blowen also received an e-mail from a woman in Hollywood, Wendy Averill, who wanted to make a donation and offered to help any way she could. Blowen told Averill that he had sent out certificates to California-based Jack Van Berg, trainer of 1988 Horse of the Year Alysheba, but he had not yet responded. Averill was able to get Van Berg to participate, plus she contacted Peggy Whittingham, the widow of Ferdinand's Racing Hall of Fame trainer, Charlie Whittingham. Thanks to Averill, Blowen scheduled a fund-raiser on April 17, 2004, to honor Ferdinand, Narrow Escape, Peggy Whittingham, and Penny Chenery, Secretariat's owner and an equine activist. The fund-raiser drew five hundred people and raised more than $12,000.

Old Friends founder Michael Blowen with Rich In Dallas, who played Seabiscuit.

About a month earlier, Blowen received an e-mail from Cathy Riccio, who said one of the horses who portrayed Seabiscuit in the wildly successful film adaptation of Laura Hillenbrand's best-selling book about the legendary champion was racing in a $2,500 claiming race at Los Alamitos in California. "I said, 'I don't care what we do, we have to claim him,' " Blowen said.

ReRun and the Exceller Fund raised $2,500 to claim the horse. A kind-hearted California trainer who wished to remain anonymous claimed the horse for Old Friends and donated him to the program. Then, Tex Sutton Equine Airways donated the airfare to ship the horse, a nine-year-old gelding named Rich in Dallas, to Kentucky.

When Rich in Dallas arrived at Blue Grass Airport in Lexington, there were TV cameras and newspaper reporters ready to greet this modest Arkansas-bred who had portrayed an equine immortal. "He didn't have to be fast; he had to just look fast," Blowen noted.

Thus, with much fanfare, did Rich in Dallas join Narrow Escape at Old Friends. "I look at both these horses as proving our point," Blowen said. He continued:

> If you have a really good horse and a really good story, people will get excited about it, and it doesn't have anything to do with money. All these horses have tremendous stories. It just takes someone to take the time to write them.

Thanks to Blowen, a lot of people are reading a lot of interesting stories about horses today. People now know that Strike the Gold and Sea Hero are in Turkey, Breeders' Cup Classic winner Proud Truth is in Panama, Belmont Stakes winner Colonial Affair is in Argentina, and Kentucky Derby and Preakness Stakes winners Charismatic and War Emblem are in Japan. Unfortunately, Belmont Stakes and Travers Stakes winner Temperence Hill died in Thailand and champion turf male Manila died in Japan before there was an Old Friends to rescue them.

"It's unbelievable," Blowen said. "People love these horses. These are fans, not bettors. It's unbelievable that when these horses are done breeding, they go to so-called riding academies, which is a euphemism for a slaughterhouse."

Not any more.

"The two most neglected groups in racing are the horses and the fans," Blowen said. "We're trying to make the connection."

CHAPTER 17

Saving Standardbreds

Sergeant York, a Standardbred who raced at Freehold Raceway in New Jersey and at Yonkers Raceway just north of New York City, could have been sold to slaughter following his unremarkable career when he competed under the name of Allaboard Jules. Foaled on April 25, 1991, Allaboard Jules won five of twenty-three starts from 1994 through 1996, earning $14,881. His final race was at Freehold on January 25, 1996, when he finished third in a $4,000 claiming race for trainer David Brandwine.

Who could ever have imagined that Allaboard Jules would have a second career—and a very public one—after the finish line? An entire nation watched Allaboard Jules, under a different name, follow the casket of former President Ronald Reagan on June 11, 2004, in the funeral procession. Sergeant York was riderless, continuing a long-standing tradition of saluting a fallen hero.

The tradition dates back to Genghis Khan and was a welcome alternative to the previous custom of sacrificing the horse of a slain warrior so his dead owner's spirit would not have to walk once he crossed into the afterlife. The sacrificed horse wore a hood covering his head, armored covering, and a saddle with the stirrups inverted and a sword through them. The riderless horse is referred to as caparisoned in reference to his ornamental covering.

Abraham Lincoln was the first United States president honored by the inclusion of a caparisoned horse in his funeral. When Lincoln's body was taken from the White House to lie in state at the Capitol rotunda, the casket was followed by Lincoln's horse, with his owner's boots backward in the stirrups.

In 1963, following the assassination of President John F. Kennedy, the role of riderless horse went to Black Jack, the last horse issued to the U.S. Army and the last horse to carry the "U.S." brand. Black Jack, named for General John J. "Black Jack" Pershing, was foaled on January 19, 1947, but his breeding was unknown. Black Jack also performed in the funerals of Presidents Herbert C. Hoover and Lyndon B. Johnson before he was retired on June 1, 1973. He died on February 6, 1976, at age twenty-nine. His ashes were placed in an urn at his monument in Fort Myer in Virginia.

Allaboard Jules seemed an unlikely candidate to replace him, but Brandwine realized the horse's gray and black colorings with no marking—a prerequisite to be the riderless horse in a president's funeral and other ceremonial processions—as well as his temperament might qualify him for the honor. Marie Dobrisky, an employee of the New Jersey Racing Commission at Freehold, knew through her son, Sgt. Frank Dobrisky of the Caisson Platoon, that the Army was looking for that type of horse. Frank Dobrisky said:

> I'd told her to be on the lookout for horses for us. The major requirement is that they have to be very sound. We only use gray and black horses with no markings, although many of the blacks we have are really dark bays. Most of the horses we use are big. We have English Shires and can put out a whole team of black or gray Shires. But the lead pair in a six-horse caisson team are lighter, smaller horses.

Marie Dobrisky called her son to tell him that Allaboard Jules might fit the requirements. Frank told his mother that he needed grays, so Allaboard Jules was not invited to the unit for nearly a year before arriving at Fort Myer in Arlington, Virginia, in the summer of 1997.

"The guys made fun of him," said Frank Dobrisky, who has done more than seven hundred missions. "They made fun of his name, and he was small and scrawny compared to our other horses. He had to have a new name, so we gave him a military name."

Dobrisky chose the name Sergeant York, for the World War I hero who single-handedly captured an entire enemy platoon.

A few weeks into training and under his new name, Sergeant York was involved in routine infantry exercises. Said Dobrisky:

Sergeant York, a Standardbred formerly named Allaboard Jules, was the riderless horse in President Ronald Reagan's funeral procession.

He was part of the opposing force in the exercise. They were simulating a Third World country that used horse-mounted troops. He did excellent. He had machine guns fired from his back, had grenades blowing up beside him. But he did a great job.

That performance helped earn Sergeant York a promotion in his second career. He was made the understudy of the riderless horse. When that horse, who had taken over from Black Jack, went lame twenty-four hours before a mission scheduled for November 7, 1997, Sergeant York was pressed into service.

"He had put some weight on, and he muscled up nicely, but he had to do this job the next day," Dobrisky said. "On the drop of a dime, he had his first mission, and I had the honor of leading him that day. There were two missions that day and he did as well as any riderless horse I've walked."

In 1998, the riderless horse that Sergeant York had subbed for died, and this former claimer was promoted to the top spot. He has aced mission after mission ever since.

"The word is out—Sergeant York is not to be ridden," Dobrisky said. "We don't want to take a chance that he'll get sore or lame. He's a very popular horse. People come into the barn [which is open daily for tours] just to see him."

Perhaps the immense attention Sergeant York has received—including a story in *Sports Illustrated*—could be enough to prompt the United States Trotting Association, the registrar of Standardbred racing and breeding, to change its position against the American Horse Slaughter Prevention Act. It is already rethinking it.

When the American Horse Council issued its white paper in 2002, calling slaughter a necessary evil, the United States Trotting Association endorsed the position. "Nobody is really in favor of slaughter," USTA Executive Vice President Fred Noe said at the time. Now, the USTA is reassessing its position. "Yes, we are," Noe said in April 2004. "This has come up at a board meeting and we are looking at our position. I would never have any of my horses slaughtered."

In the time from 2002 through 2004, the U.S. Trotting Association supported programs that provided options to slaughter. "They've been very supportive of our work," said Judy Bokman, who co-founded the Standardbred Retirement Foundation in 1989, with Paula Campbell, the wife of harness racing's all-time leading money-winning driver, John Campbell. "When we need help, Mr. Noe has helped us in any way he can. Do they give us $100,000? No. Do they give us $10,000? No. But they help us."

Bokman has been around horses for much of her life. She recounted:

> I got involved in harness racing because my uncle, Anthony Chiarello, and I owned a horse, Nadala. I still have him. He's thirty-three. I had him up at Goshen, New York [home of the Harness Racing Museum and Hall of Fame sixty-five miles northwest of New York City], for about two months and I witnessed the Amish coming to buy the horses. They pulled up to the track, and they'd buy them. I looked into it further and I saw what was happening about twenty years ago.
>
> The Amish are businesspeople. They use these horses and most of them go to slaughter. I tried to find a few people who would take a horse to adopt. Then Paula and I put together the foundation. It's a terrible injustice that we're using horses for monetary gain, not only for an individual, but for an industry, and there are no provisions for these animals. In this huge industry, there's no reason something else can't be done.

When they can no longer make money on a racetrack, many Standardbreds are sold to the Amish, which is a better option than slaughter but hardly a ticket to paradise. "You can call the Amish and sell a horse for $250 or $300, which really isn't the answer," Paula Campbell said. "They work to death for the Amish. When the Amish get done with them and they can't walk anymore, they'll send them to slaughter. We're prolonging their life, but it's not the answer. They're not getting a better life."

With the Standardbred Retirement Foundation, they do. In mid-March 2004, the foundation was paying to board 125 horses. All but fifty are so badly injured that Bokman terms them "completely nonusable" for a second career. "They have chronic injuries," said Bokman, who lives on a small farm in Monmouth County, New Jersey, with her veterinarian husband, Steve, and their two sons.

The Standardbred Retirement Foundation has saved and placed more than 1,350 horses in adoptive homes since its inception. "Anybody that's been involved with horses and works with them can't be in favor of horse slaughter," John Campbell said. "They [the Standardbred Retirement Foundation] are in it for the long haul. It sheds a good light on the whole industry."

His wife sees no other options. "We'd like to give the horses some dignity when they're done racing," Paula Campbell said. "That's what this industry is all about, the horses. If we can't save them, that's pretty sad."

As a result of Bokman's and Campbell's passion, the Standardbred Retirement Foundation has won the Association of Racing Commissioners International's Animal Welfare Award in 1995, the Proximity Award in 1997 from the United States Harness Writers Association, and was a finalist for the American Association of Equine Practitioners's Lavin Cup.

Besides never turning down a horse, the foundation follows up on each adoption. If the adopted horse's new owners can no longer care for the horse, the horse is taken back into the foundation's care. Early in 2003, foundation veteran horses Valentine and HT's Magnum were returned, but not for long. Both horses were quickly adopted by the mounted police forces of Hoboken and Edison, New Jersey.

The breadth of the Standardbred Retirement Foundation's impact is considerable. In 2003, other foundation success stories included a twenty-three-year-old veteran, a two-year-old trotter, and a horse building up a lot of frequent flier miles. Archie, a twenty-three-year-old pacer with a tattoo too faded to be read, was retired as a pleasure horse with Becky Fara in Blairstown, New Jersey, where he joined Baby, a twenty-three -year-old who had been placed at the same farm a year earlier. Cavalier Hall, a three-year-old trotter, was placed with the foundation after a barn accident caused him partial blindness in one eye, effectively ending his racing career. Diane Alberts of Howell, New Jersey, adopted Cavalier Hall and they have embarked on a new career in dressage. Canaco Jackomo won the Standardbred Retirement Foundation's award for traveling the farthest. He was adopted by Chrissy Roness of Fort Rucker, Alabama, and moved with Roness's military family to Fairchild Air Force Base in the state of Washington.

Probably, the most famous harness horse the Standardbred Retirement Foundation has helped is 1981 Hambletonian winner Shiaway St. Pat. Once at the top of the sport when he won harness racing's premier trotting event as a three-year-old, he eventually dropped down to claiming races, where any licensed owner could buy him for a specified price. "He was eleven or twelve and had been racing in cheap claimers," Bokman said. "It was shameful. Everyone who knew the horse felt that way. It was a disgrace. He won a million-dollar race. He was racing even though he had a pile of injuries."

The Standardbred Retirement Foundation and the Meadowlands, the showcase track of North American harness racing and home of the Hambletonian in East Rutherford, New Jersey—not far from the Standardbred Retirement Foundation's national office in Freehold, New Jersey—came to Shiaway St. Pat's rescue.

"The Meadowlands got him and they put him out in Paddock Park for Hambletonian Day, and we found him a home with a veterinarian," Bokman said. "We try to ensure that the new owner is a responsible, experienced horseperson. There's a simple, four-page application form people must fill out, and we ask for references from a veterinarian, a farrier, and a friend with horse experience. We also reserve the right to reclaim any horse who is not receiving proper care. We're looking for lifetime homes for these horses. So, as part of a contract the new owner signs, the horse may not be sold, raced, employed for any commercial use, or bred."

The foundation does not help just horses. It also reaches out to at-risk, physically and mentally challenged children through its "Youth in Focus" program. The foundation works with school counselors to target students most likely to benefit from spending after-school hours or parts or all of their summer working with horses. With guidance, two students take on the responsibility of caring for a single horse, learning to work together, not only with each other but with the horse. In doing so, the child's self-esteem is boosted. This is heightened even more when the two children and the horse reach their goal—finding a suitable, adoptive home for the animal. "It is amazing to see the magic that these wonderful horses can do for these youngsters to give them a better perspective on life and a meaningful purpose," Bokman said.

The scope of the Standardbred Retirement Foundation continues to grow thanks to the commitment of key players in the industry determined to change the status quo of horse slaughter in America.

On July 28, 2004, the Meadowlands initiated the Standardbred Retirement Foundation Race for three-year-old pacers, with two separate divisions. On October 28, 2004, Freehold Raceway held the Standardbred Retirement Foundation race for two-year-old colt and gelding pacers. A horse is eligible for the races if a breeding to that horse's sire is donated to the Standardbred Retirement Foundation. A donation, which is resold commercially to raise funds, makes all of that sire's two- and three-year-olds eligible for the races. Initiated in 1994, Standardbred Retirement Foundation races have generated more than one million dollars in purse money to the winning connections. Approximately one hundred stallions are eligible every year.

In the summer of 2002, when the U.S. Trotting Association accepted the American Horse Council's position that slaughter was a "necessary evil," Russell Williams of Hanover Shoe Farms in Hanover, Pennsylvania, took the opposite position, even though he is vice chairman of the board of the U.S. Trotting Association and on the board of trustees of the American Horse Council. In a letter to Christopher D'Arcy, the staff director of the livestock and horticulture subcommittee of the U.S. House of Representatives Agriculture Committee, Williams noted that Hanover Shoe Farms was founded in 1926 and remains the largest Standardbred breeding farm in the world. In 2001, Hanover Shoe Farms bred horses that won more than $21 million in purses, an all-time record for any breed. Williams wrote,

> Many of the worst abuses occur routinely in the sale, transport, and 'processing' of horses slaughtered for meat. Here at the Hanover Shoe Farms, we retire our horses that have reached the end of their productive lives and continue to take care of them until they either die of natural causes or must be humanely put to sleep. We would not consider relegating them to a bad end after all they have done for us. Currently we have fifty-two retired horses on the farm.

> Discussion of this legislation [to ban slaughter] has already raised the consciousness of many members of the American horse industry, myself included, of the need to alleviate unnecessary suffering among horses, which are entirely dependent upon us, and our legislators, for their welfare. In my estimation, this trend is growing. Thank you for your attention in this important humane issue.

You do not have to be an active participant in harness racing to be humane to its horses. Jack Chellew, the senior sales representative in the Equine Group for Merial Pharmaceuticals, won the second annual Standardbred Retirement Foundation's Burke Humanity Award, named for the late Standardbred owner and trainer James E. Burke, in 2003. Chellew conceived and implemented a program to distribute soon-to-expire but still-effective drugs that otherwise would be disposed of by their manufacturers after the drugs had passed their sale date. Now, a variety of equine charities are acquiring drugs they might not otherwise be able to afford through their local veterinarians.

But the people who save Standardbreds want more than restoring their horses to health. They want to find them second careers. And Standardbreds are well suited to doing that, as the Standardbred Retirement Foundation explains on its website "Imagine a horse which every family member can enjoy, a horse which is kind, gentle, versatile, and more" (see box, p. 145).

"People tend not to realize the potential and versatility of the Standardbred breed," said Karen Beach, former administrator of the Standardbred Equestrian Program, which was launched in November 1995. "The Standardbred Equine Program was developed in an effort to create the option of a second career for these horses. We wanted to take a visionary stand, instead of a reactionary one."

Another one of its programs is CHAMP (Care, Horsemanship, Achievement, Merit, and Promotion), which encourages young people to work with Standardbreds.

"Standardbreds can make terrific saddle horses," Beach said. "They are already familiar with the feeling of girths around their bellies, verbal cues, and hands on the reins. They only need to learn to deal with weight on their back and leg signals to make the transition from the track to a pleasure horse career. The Standardbred's natural stamina is also making the breed an increasingly popular choice for endurance riding."

Horse Lovers United Inc. has been finding new homes and new careers for all breeds, while specializing in retired or noncompetitive

Why Adopt a Standardbred?

Imagine a horse whose disciplined track life has gifted him with qualities most desired in the world of pleasure riding. The horse is a Standardbred, a cross between the Thoroughbred and the Morgan. These gentle, beautiful horses are available through the foundation's award-winning adoption program. Standardbreds are making the successful transition to lives and careers that are far removed from the racetrack. The Standardbred horse has a great deal of versatility and has so much to offer. In addition, these beautiful horses are able to face every task put before them with gentleness, patience and a heart that knows no limit.

The Standardbred has already been trained to the bit, and knows how to walk, trot, stop, back up, and stand patiently in crossties. Standardbreds also enjoy being groomed, bathed, and fussed with. Additionally, they know how to load in a horse trailer, travel, and unload. Many Standardbreds have experience around noisy, horsy, active places like a race paddock and/or racetrack. Lastly, these horses clip very well, have had their legs wrapped, and usually have been taught to wear a blanket and a cooler.

STANDARDBRED RETIREMENT FOUNDATION www.adoptahorse.org

Standardbreds, since its inception as a not-for-profit organization in August 1992. The organization was founded by a small group of dedicated horse lovers whose goal is to find new, lifetime homes for displaced horses.

Horse Lovers United has placed more than 110 horses, who have found new careers in dressage, trail riding, pleasure riding, riding in parades, and sleigh riding. Horse Lovers United has a main office in Salisbury, Maryland, a chapter on the eastern shore of Virginia, and two phone contact numbers in Delaware.

All donated horses start with a transition from working horse to pleasure horse. The horse is boarded at a private farm, evaluated by experienced horsemen, given necessary veterinary care, groomed, and started under saddle to determine if they are capable of a new career in riding. Horses are then matched with applicants, who are carefully screened and must sign an adoption agreement. Follow-up visits are made periodically to ensure the adopted horse is receiving quality care.

If an adopter cannot continue to care for a horse, that horse is readmitted into the program.

Saving Dutchess

It is hard to describe a blind horse as lucky, but Dutchess, a ten-year-old Standardbred mare, was. She was saved from killer buyers twice at the Sugarcreek Auction in Sugarcreek, Ohio, finally finding a home with Annette Fisher's not-for-profit Happy Trails Farm Animal Sanctuary in Ravenna, Ohio.

Fisher said they had to turn down Dutchess several times because Happy Trails was overflowing with rescued horses. She eventually moved around some horses, freeing up a stall, and Dutchess was brought to Happy Trails.

Fisher, a native of Barberton near Akron, Ohio, is adept at rescuing horses and finding them homes. She estimates she has placed sixty horses in the past three years, a majority of them Standardbreds and draft horses. But the first horse she took in was a twelve-year-old Thoroughbred mare named Real North, who had been racing at Thistledown racetrack in Cleveland. Fisher and her husband, Russ, a long-distance truck driver, had just moved into rural Ravenna, and Fisher saw it as an opportunity to follow through on a childhood passion.

"It was a dream of mine," Annette Fisher said. "Ever since I was little, I wanted a horse."

She bought Real North in 1998 for $850, hoping she could turn her into a backyard riding horse. "I boarded her while our barn was being fixed," Fisher said. "But you can't have just one horse. They need a buddy."

So Fisher bought a PMU foal from Canada, a three-month-old Belgian she named Shelby. Shelby, not Real North, became Fisher's riding horse.

But it was a pig who, in Fisher's words, kicked Happy Trails "into high gear." Fisher was talking with a woman from a nearby farm, discussing how difficult it was to take a vacation if you own a farm. Fisher offered to watch the woman's farm while she took some time off. When she visited the woman's farm, she was led to a barn and the woman opened a stall door. "It was a horrible little dungeon, a four- or five-foot-square area," Fisher said. "In it was a potbellied pig who was lying in its own feces. I said, 'What's wrong with the pig?' She said, 'Somebody dropped it off six years ago, and I think it had two broken legs.'

"This picture haunted me," Fisher said. "When the woman came back, she asked, 'What do I owe you for watching the farm?' I said, 'How about I take the crippled pig?' "

Thus did Janice the pig—whose two legs were not broken but had atrophied—find a new home, where she still lives. "She'll never walk again," Fisher said. "She was completely crippled, but she's doing well.

We built her a log cabin. And then we had to get another pig for a companion, George."

When Fisher's husband noted, "You can't keep bringing these animals home," Fisher did some research and created Happy Trails as a not-for-profit charity.

Which is how Happy Trails became home to twelve horses, a cow, six farm hogs, twenty-one potbellied pigs, four dogs, six barn cats, and a mixture of geese, chickens, and turkeys in mid-April 2004. But Dutchess was Fisher's biggest challenge.

Dutchess suffered from uveitis, an inflammation of the globe, or anterior chamber, of the eye. "Uveitis is a horrible disease," Fisher said. "It not only causes complete blindness but also causes the eye to shrink back in the socket. The horse can't close her eyes, and debris, dust, and flies get in, which results in painful eye infections."

Eventually, the eye tries to heal itself and scars, and the globe shrinks within the socket, which can cause chronic damage.

Most horses get uveitis in one eye, but Dutchess had it in both. Veterinarians recommended removing both eyes to relieve the pain and to minimize the chance of brain damage.

On February 2, 2004, Caroline Tessier, D.V.M., at the Aurora Equine Specialty Hospital in Mantua, Ohio, removed Dutchess's left eye. A global prosthesis of composite plastic was inserted for a better cosmetic appearance. A second operation would be performed later.

"This will help her overall appearance and chances for adoption," Fisher said. "We would love her no matter how she looked, but we are hoping to adopt her to a loving home."

Dutchess checks out the corners of her stall with her nose and whinnies at Corona, a fifteen-year-old anemic Standardbred mare in an adjoining stall who is recuperating from the effects of malnutrition. Other equine residents at Happy Trails in the spring of 2004 included Duesie, an eighteen-year-old Standardbred gelding, Sequoia, a seventeen-year-old mare who is blind in one eye, and Rocky, a ten-month-old Percheron colt who was recovering from neglect and malnutrition.

On April 11, 2004, after Dutchess had learned how to be a riding horse, she was adopted by Joy Wagner, an equine massage therapist, a Happy Trails volunteer, and its attorney. "Dutchess is doing wonderfully," Fisher said.

Dutchess was the third horse Fisher placed in two weeks, following Evita, a ten-year-old chestnut Arabian, and a Hackney. Four other adoptions were pending. The little girl who dreamed of owning a single horse has saved the lives of many.

She Saves Hundreds

Dot Morgan is a seventh-generation horsewoman from Louisville, Kentucky, who used to ride in the annual Kentucky Derby parade. When she and her husband, Charlie, a harness trainer and driver, settled in Ohio, she knew there had to be a better way to take care of racehorses after their careers were over. A conversation with a truck driver who hauled horses to slaughterhouse auctions changed her life. It also saved the lives of hundreds of horses.

"I saw these horses going to the stockyards," said Morgan, whose nephew Tony is one of the leading harness drivers in North America. "I knew one of the truckers, Charlie, who hauled the horses off. He lived in Dry Ridge, Kentucky. I had a cup of coffee with him one day and I said, 'Charlie, all these horses you're shipping to the stockyards, are they all injured?' He said, 'No, but they're crazy. By the time I get to the auction, half of them are on the floor of the trailer and the other half are trampling them.' "

It was a deeply disturbing image she could not shake, one foreign to her life as a horsewoman and to the role horses had played in the raising of her two daughters, Anna and Winnie, who both became equestrian champions. "It was a terrible scenario," Dot said. "I just made a decision to do something about it. What was lacking was someone to stand in the gap who knew horses."

She knew who could fill that gap. All she had to was look in the mirror.

Her New Vocations Racehorse Adoption Program began in 1992, when she placed a small ad in the *Thoroughbred Times* saying she would take unwanted horses and find homes for them. One of her first respondents was John Franks, a leading Thoroughbred owner and breeder who died December 31, 2003.

Dot Morgan explained:

> New Vocations started as a mission, not a charity. We took twenty to thirty horses and it wasn't hard finding homes for them. But they weren't for beginners. The horses needed retraining. They needed to have a transitional course in becoming a pleasure horse. It didn't take long, but it takes a patient person to help them make that transition. I train them myself.

She has help from her twenty-five-year-old daughter, Anna Ford, a World Champion Saddle Seat rider who became the New Vocations Adoption Manager. Ford works at the Sid Griffith Equestrian Center in

Hilliard, Ohio, about fifteen miles west of Columbus, where New Vocations leases paddocks and stables. The other New Vocations site is in Laura, Ohio, north of Dayton, where there are thirty-two acres of rolling, creek-fed pastures, paddocks, and barns. Morgan's other daughter, Winnie, a National Champion Stock Seat rider, boards horses for New Vocations near her home in Ann Arbor, Michigan.

New Vocations' program grew quickly. "It grew by leaps and bounds," Morgan said. "It stayed small until 1998, maybe forty horses a year."

But, acting on a suggestion by Thoroughbred breeder and owner Paul Robsham, New Vocations became a 501(c)(3) charity. The number of adoptions has boomed since, reaching 239 in 2003.

This has been accomplished on an annual budget of only $250,000.

In 2003, New Vocations received $64,701 from donations and fund-raisers and continued to benefit from the Morgan Perpetuity Endowment Fund, a $50,000 grant from a prominent Thoroughbred breeder. A Standardbred stallion auction raised $43,745. Blue Horse Charities contributed $19,650, and Thoroughbred Charities of America provided a $20,000 grant.

That support helped New Vocations in 2003 place 239 horses, 166 Thoroughbreds and seventy-three Standardbreds, from eighteen states: Arizona, Florida, Illinois, Indiana, Kentucky, Louisiana, Massachusetts, Maryland, Michigan, New Hampshire, New Jersey, New York, Ohio (which had 138), Pennsylvania, Tennessee, Vermont, Washington, and West Virginia. The horses from those states were then placed in twenty-nine different states: Alabama, Connecticut, Florida, Georgia, Iowa, Indiana, Kansas, Kentucky, Louisiana, Maryland, Michigan, Minnesota, Mississippi, Montana, North Carolina, North Dakota, Nebraska, New Jersey, New York, Ohio (high with eighty-seven), Pennsylvania, Rhode Island, South Carolina, Tennessee, Texas, Virginia, Vermont, West Virginia, and Wisconsin.

Making those numbers even more remarkable is the strict adoption program Morgan has in place. Only a previous horse owner can adopt a horse, and that means filling out a fifty-one-question application form, having one reference from a veterinarian, plus two other references; and then, once the adopter signs a two-year contract, sending back to New Vocations six-month reports with pictures of the horse. Morgan and her daughter do that with the aid of two part-time workers and four or five college interns from Ohio State University, Otterbein College, and Wright State University.

In addition, New Vocations now works with girls in the West Central Ohio Juvenile Center, yet another example of once-doomed horses helping change the lives of people.

Since 1992, New Vocations has found new homes and new lives for more than one thousand Standardbreds and Thoroughbreds. “We’ll do more,” Morgan said. “Our goal is to get as many through here as we can.”

CHAPTER 18

A Breed Apart

While there are more than two hundred programs that rescue, rehabilitate, and retrain Thoroughbreds and Standardbreds to save them from slaughter, what about the other major breed of American racehorse, the Quarter Horse, the most popular breed in America? Who protects them?

The number of Quarter Horses in the United States is staggering. In 2003, according to the American Quarter Horse Association (AQHA), there were more than 2.8 million Quarter Horses, including 474,000 in the state of Texas alone. That represents 40 percent of the total national equine population of 6.9 million estimated by the American Horse Council. The 2003 Thoroughbred population was estimated to be 725,000 by the Jockey Club. The United States Trotting Association estimated there were 80,000 to 90,000 Standardbreds in the United States in 2003. That year, there were more than 141,000 new Quarter Horses registered in North America. By contrast, the 2003 Thoroughbred foal crop registered with the Jockey Club totaled 37,000. In terms of racing, there were 14,532 individual Quarter Horse starters in 2003. That same year, there were 73,614 Thoroughbred starters in North America.

When the American Horse Council published a 2002 white paper calling slaughter a "necessary evil," the American Quarter Horse Association substituted its name for the American Horse Council's and sent out the white paper as its own position statement. But while the American Horse Council quickly backed off its position—it now claims to be officially neutral on slaughter—the American Quarter Horse Association has maintained its slaughter-is-okay posture. More disturbingly, the American Quarter Horse Association joined the American Association of Equine Practitioners and American Veterinary Medical Association in lobbying against the anti-horse slaughter bill before both houses of the United States Congress.

That is why Blue Horse Charities founder John Hettinger circulated a letter throughout the racing industry in November 2003 that chastised the American Association of Equine Practitioners, American Veterinary Medical Association, and American Quarter Horse Association for their role in keeping the anti-slaughter bill bottled up in the U.S. House of Representatives Agriculture Committee, where it has resided despite the fact that it has more than 220 cosponsors.

The American Quarter Horse Association responded with its own industrywide letter, saying the Blue Horse Charities letter had "several misleading, incomplete, and false claims" about the AQHA. Ward Stutz, the AQHA director of member services and public policy, wrote in the letter:

> AQHA does not support [the legislation], and the association does not favor slaughter as a means of dealing with unwanted horses. What AQHA favors is horse owners having all legal choices available for the difficult end-of-life decisions. . . . That said, AQHA does have serious concerns about any federally imposed ban on processing facilities. Chief among our concerns is taking away a viable humane euthanasia option, which would increase instances of abuse and neglect of unwanted horses. This has been supported by an independent scientific study. We do not want to see horses starved, mistreated, neglected or shipped to other countries where processing facilities are not regulated by U.S. law or USDA [U.S. Department of Agriculture] officials. AQHA also recognizes that adoption and retirement facilities are options for owners, and would endorse federal regulations and standards that would govern the operations of these facilities.

What retirement/retraining/rehabilitation facilities are out there for Quarter Horses? "There's a number of them, but we don't endorse any single one of them," Tom Persechino, the AQHA senior director of marketing, said in April 2004. "We recognize different programs. A lot of our horses off the racetrack go on to second careers, including barrel racing, roping, and jumping. There's not one particular program for Quarter Horses. One of the things we'd like to see is regulations which would govern those businesses [*sic*, Most are 501[c][3] charities.—Ed.].

There's no legislation about the care of horses there. We'd like to see addressed a set of guidelines."

To the AQHA, then, the people who have volunteered their time, and in many cases, devoted their lives, to saving horses must be running their programs in a manner that offends the AQHA, and that they are not properly taking care of the horses.

"As far as the issue of horse slaughter is concerned, the AQHA does not favor horse slaughter," Persechino said. "We recognize it as an option. At the same time, we do oppose any legislation which would ban the right of people to manage their horses."

To the AQHA, "managing" a horse does not require personal responsibility to care for the horse after its use as a riding horse or racehorse has ended, but rather the right to sell the horse to a slaughterhouse.

"We've been to a slaughterhouse and seen it firsthand," Persechino said. "We do think of it as a humane way to deal with unwanted horses. I do happen to believe that without that option, horses would suffer a crueler fate. The last thing anyone wants is to see a horse abandoned or neglected."

So where, then, are those 300,000 or so less horses slaughtered in the United States from 1989 through 2003? Where are all the stories about neglected and abandoned horses? And why has the number of abused horses diminished in California after the slaughter of horses was banned in 1998?

They are safe with all the people who are devoted to providing productive lives to horses after the finish line.

CHAPTER 19

The Washington Wonder

While the two principal veterinarian organizations in the United States, the American Association of Equine Practitioners and the American Veterinary Medical Association, warn of thousands of neglected and abandoned horses roaming the countryside if the slaughter of horses for human consumption is abolished in America, the reality is that a single neglected horse in 2003 drew national media attention. And it was not in the horse country of Kentucky or near a major city, but in the rural setting of Lake Stevens in Snohomish County, forty miles northeast of Seattle in the state of Washington.

The mystery horse was found eating grass on the side of a road in Lake Stevens shortly before Christmas. He was dragging a lead rope and wearing a halter. Neighbors thought the horse belonged to the family of Steve and Erin Porter, who board horses on their Lake Stevens farm. Erin, a forty-year-old waitress, rides and has been around horses her entire life. Steve, forty-seven, is a network technician.

"I came home and the horse was in my pasture," Erin Porter said. "There was a note on my post by the driveway that the horse had been found and they thought it might be ours. When I called Animal Control, I told them the horse didn't belong to us."

The Porters agreed to care for the horse temporarily at their farm. The horse had ringbone, a bony growth of his front hoof that caused him to stumble, but other than that and a nasty gash on his nose, he was physically healthy. "I hate to see somebody discard an animal and not take care of it," Steve Porter said. "It just doesn't make sense to me."

Though everyone involved thought the mystery horse was a Quarter Horse, Erin thought the horse was a Thoroughbred. She could have verified her feeling if she had been able to check for the blue tattoo every registered Thoroughbred receives inside his upper lip. The number is recorded by the Jockey Club, the breed registry.

"He wouldn't let me close enough to him to lift up his lip," Erin said. "It was evident he had been through trauma. He had a big gash on his nose and he wouldn't let anybody near him."

The Porters assumed the horse's owners would realize he was missing and contact them, but, despite widespread publicity, no one did. Said Erin:

> I put flyers up and nobody ever claimed him. I had him for about a month, and they took him to the horse auction. They had told us we'd be able to keep him because we were working so hard to find the owner. The city wanted money for him. Had it been a dog, I would've been able to keep him.

But sending him to auction allowed someone there to finally discover that he did have a tattoo number. That allowed him to be identified. His name was Flying Algonquin, a dark bay Thoroughbred gelding who had won one of six starts and earned $8,160 racing at Hastings Park in Vancouver, British Columbia.

With the assistance of Ralph Vacca, the general manager of the Washington Thoroughbred Breeders Association, the Porters were able to trace the horse's history. Flying Algonquin was a son of Vying Victor and was foaled on April 17, 1997. He had been bred and owned by Fred and Elizabeth Brandt of Aldergrove, British Columbia, who nicknamed him "Stretch" at birth. Following his three-year racing career, the Brandts sold Flying Algonquin for $2,000 as a riding horse to a couple in Snohomish County. Flying Algonquin then changed hands at least two more times.

Despite that information, as an impounded animal, Flying Algonquin had to be sold at the nearest livestock auction, which was in nearby Marysville. But Flying Algonquin, whose story would be picked up and run nationally by the Associated Press, had a lot of friends he would never meet. Calls came in from all over the country. Then, by e-mail and word of mouth, two local 4-H clubs in King and Snohomish Counties raised more than $1,300 to give the Porter family a chance to buy him and prevent him from being sold at the Marysville Livestock Auction on February 3, 2004.

"I got so many e-mails," Erin said. "Seventy people from all over the country donated money." By then, the Porters had failed miserably at their attempt to remain detached from Flying Algonquin.

Though the auctioneer announced Flying Algonquin might be unable to serve as a riding horse because of his ringbone, the bidding on the horse climbed past the $1,000 mark. Steve Porter bid $1,375 and the lone person bidding against him, a man in a cowboy hat, shook his head no, meaning that he was done bidding. The auctioneer yelled, "Sold!"

"I was so relieved," Erin said. "I knew his future was secure with us. I didn't want him to go to slaughter. Without all the money donated, we wouldn't have been able to bid that high. I was extremely grateful to all the people who helped us."

The Porters will ask Flying Algonquin to do absolutely nothing for the rest of his life, and he will live in peace at the Porters' farm in Lake Stevens. Three months after being rescued from slaughter, Flying Algonquin had mellowed.

"He's a wonderful horse," Erin said. "I've been working with him every day. I pet him and show him we love him. I went out today to take his fly mask off him and he laid his head on my shoulder."

The bond of horse and human had healed.

So Many Others

While the American Association of Equine Practitioners and the American Veterinary Medical Association argue that the brutal slaughter of horses is a necessity because horses would be worse off without that option, more than two hundred organizations in the United States are proving the opposite, and it seems as if there is a new program starting every week.

The slaughterhouse deaths of Exceller and Ferdinand accelerated a movement that continues to grow as more people learn about horse slaughter. Unlike the veterinary organizations charged with preserving the welfare of horses, these organizations are doing something about it. "There's a lot of horses out there who don't get taken care of," said Steve Porter of Washington.

Through the end of April 2004, these 200-plus programs that save horses from slaughter were located in forty different states. The state of California had at least twenty-two separate programs; Pennsylvania at least fourteen; New York at least twelve; Florida, Maryland, and Texas at least eleven each; Colorado at least nine; and Arizona, Michigan, and Ohio at least eight apiece. That was as of the summer of 2004. And those were the public programs. There are others out there.

Horses are being saved from coast to coast, and the slaughter of horses in America for human consumption overseas inevitably will end. The only variables are how long it will take and who will help end it.

CHAPTER 20

From London to Heaven

Seventy years ago, when he five and growing up in London, Peter Gregory would watch the horse-drawn dairy carts depart from and return to a nearby milk plant. "I'd watch the carts go out every morning," he said. "When they came back at night, the horses had had a heavy day. I used to walk down to the dairy. They had to unload the empties. And the horses were taken to their stalls. Those horses, basically, never saw any green fields to graze on. I thought when they sleep, they must dream of green pastures. They had a hard life.

He came to the same conclusion a few years later watching horse-drawn wagons leaving a bakery. "They used to pull out every day with heavy loads," he said.

When he met Mary, his wife of fifty years, in a classroom at the University of London, he found a kindred spirit. "We both had a love of animals, not just horses," he said. "It's rare to have a couple like the same things. We think the same."

In the early 1950s, they would visit an equine sanctuary near Elstree, where carriage and cart horses would be given a two-week break from working the streets of London. "The farm gave them a two-week break from work," Peter said. "We used to take them carrots. It was the only time these horses saw a pasture. Like a thousand people before us, we said, if we ever had the cash, we would so something for these horses. Thirty-five years later, we were lucky to be able to do that."

Mary might have become a veterinarian had she been born in a different era. "She wanted to be a vet," Peter said. "Mary never got in vet school even though she was persistent. They said, 'Who would take an animal to a woman vet?' It was a different time then. It was just never done. You never heard of women doctors except in Russia. It's hard to explain to people nowadays. It was a different age. The only career a woman had in the [nineteen] forties and early fifties was secretary."

Mary, though, would eventually get a life with horses. "How beautiful they are, and how intelligent," Mary said. "I couldn't be without them. We've always liked animals and that's our life now."

It has been the Gregorys' lives for twenty years, but it took a lot of miles and a lot of years to get there. They celebrated their fiftieth wedding anniversary on May 22, 2004, but when Peter left England for Canada, Mary initially stayed behind. Peter had trouble finding work in Montreal, so he hitchhiked to Toronto. "I was basically penniless," Peter said. "When I got to Toronto, I hit the streets and tried to find a job."

He has had many in his lifetime. He worked in a grocery store warehouse and at a Distant Early Warning station in Northern Labrador. After fulfilling a six-month contract, he returned to Toronto and got a job with a movie house while attending school in the evenings. He then got a job as an accounting clerk, and eventually became an accounting executive.

The direction of his life changed when he heard about a group of investors that had purchased 102 acres in Freeport in the Bahamas for development. Using all the money he had saved, Peter went in on the deal. Eventually he became comptroller of the Grand Bahamas Development Company. "I built some duplexes and invested in a delicatessen and sold it and made a lot of money," he said. He saved a lot of money, too, because there was no income tax in the Bahamas. And he invested wisely.

Business interests took Peter, Mary, and their three children, Carol, Peter, and Paul, to Portugal, Spain, and England before Peter invested in a hotel in Jamaica. "I bought a half-interest in the place," he said. "We were there for three and a half years."

When the political climate in Jamaica changed, the Gregorys moved to South Florida and, in 1969, bought the Native Sun, a motel in Pompano Beach.

"We turned it into the first nonsmoking motel in Florida or in the U.S.," he said. "We lived there for eleven years and built up a very good clientele. We got an offer to sell it to a Canadian bank trust. They turned it into a time-sharing thing. So we had the money now."

They bought a condominium in Fort Lauderdale but rarely used it. They lived in a town house in Boca Raton and played a lot of tennis. "We had a life of leisure for a year," Peter said. "But we were always talking about doing something for animals."

At this time, their son Paul was attending the University of Florida in Gainesville. "When we came up to Gainesville, we started looking around," Peter said. "We liked the area. It was nowhere near the ocean. Gainesville, at the time, was a university town."

In early 1983, Peter and Mary saw 245-acre Mill Creek Farm in Alachua, Florida, eleven miles north of Gainesville. "Driving in, I said, 'Yeah, I can do something with this,' " Peter said. They bought Mill Creek Farm in April 1983, put their south Florida properties on the market, and built a home on the farm they would rename the Retirement Home for Horses. "We put up fences, ran water lines, and we sent the word out," Peter said. "We would take any horse that needed a home."

Their first horse was a Shetland pony named Trigger, who arrived with a heavy rope tied around his neck. When Peter cut the rope, Trigger took off. He was finally nabbed several weeks later when he was spotted on a property owned by a sheriff.

"When we got Trigger, he hated people," Peter said. "He'd eat, but he didn't want to be around people. Then, when we got another horse, Pepper, and they became good friends. Trigger became very friendly. It took two to three years. Then we got more horses from the vet school [at the University of Florida]. We took all the horses and never asked for a penny. In the first six years, we financed it ourselves. We had built barns. We were cutting hay in the fields. I realized we were going to run out of money before long."

The Gregorys decided to have an open house in December 1990 and solicit donations. Now they have an open house every Saturday from 11 a.m. to 3 p.m. The admission price to visit the horses is two carrots, though visitors are advised that a five-pound bag would be appreciated. "We asked for contributions," Peter said. "It took ten years before it covered our expenses. We have no money now, maybe $10,000 for my funeral."

National exposure in the media helped to generate the revenue stream that allowed Peter and Mary Gregory to continue taking in more horses. A story that ran in the *Jacksonville Times-Union* got picked up by the Associated Press, which ran the story nationally in 1991. "A woman rang us up from California," Peter said. "She had an Appaloosa, Harry. He had one eye." Harry suddenly had a new home 3,000 miles away.

The woman, Dannielle Dorris of Newport Beach, California, agreed to cover part of Harry's expenses at the Retirement Home for Horses by paying $40 a month as an "adoption" fee, a strategy that proved successful for the Gregorys. "She sent $40 a month every month for thirteen years," Peter said. When Peter called her up in November 2003 to tell her that Harry had died, she decided to adopt another Appaloosa there for $60 a month.

Harry and every horse who dies at the home are buried on the farm. And every time a horse dies on their farm, Peter and Mary plant an oak tree. Said Peter:

> We make two promises to a horse when he comes here. The first one is you'll never be ridden or have to work again. The second is you're going to be here forever. We bury them in what we call the 'Field of Dreams.'

The world really learned of the home when it was featured on "On the Road with Charles Kuralt." The story aired as part of *CBS Sunday Morning* on March 6, 1994. "They got over 600 letters, which CBS sent to us," Peter said. Many had sent contributions. The clip about the home was included in a twenty-five-year anniversary show.

More newspaper articles and television coverage followed. Jack Hanna of *Jack Hanna's Animal Adventures* featured the home. "He came back three years later and he said of all the several hundred shows he did, we were one of the top five that he produced," Peter said.

In the ensuing years, tennis legend Martina Navratilova and former United States Attorney General Janet Reno became sponsors and visitors. Reno adopted Bandit, the longest tenured resident at the home, and visited him on September 7, 2002, which created additional publicity. When Gainesville golfer Bill McKenzie retired from the game at age ninety-one, he donated his golf cart to the home. Peter uses it every day.

For Peter, his workday at the farm would tax anyone half his age. "He gets up at 6:30 every single morning," Mary said. "And he just loves it. He works all day and takes a break to watch the news. He has to watch the 6 o'clock news. Then we usually eat dinner. Then he'll go out and feed them, and he says it's so peaceful then, just the horses and the dogs. He'll fall asleep for an hour or so and get up. He gets up at midnight and gives the old horses their feed [a special mash because many have lost most or all of their teeth]. Then he comes back to the house and does his bills and correspondence until three in the morning. I can't do the hours. I get up at 6:30 in the morning. I have to go to sleep at nine."

In 2001, because they were getting so many horses from police forces, circuses, and handicapped riding programs, the Gregorys stopped taking horses from private individuals.

"We have eighteen police horses," Peter said. "When they're no longer able to work, the city or county auctions them off. There's only one place for them to go. That's the slaughterhouse. We feel these horses worked hard for people all those years. Unless we took them in, they'd go off to slaughter. Same thing with state park horses. There's something wrong about that. The thing that really concerns me is people who get a pet dog look after that dog until he dies. With a horse, people don't do that. People buy them on a whim when they're young. When the people get older, the horse is sold to auction."

At the home, each horse is given two acres of land to run around on. "We treat them like horses," Peter said. "They get plenty of exercise. Most of the time we leave them alone. They're not ridden and I don't want people going in there with them. I always say they're part of our family."

In 2004, the home's animal population was one hundred twenty-four horses, seventeen dogs, eight pigs, five goats, three cats, and a cow.

Among the equine residents were Wendy, a thirty-one-year-old bay mare with wires sticking out of her belly, a souvenir of her career as a research animal, and thirty-three-year-old Howie, one of five blind Appaloosas sharing a pasture. The only former racehorse the home had in 2004 was a thirty-year-old Quarter Horse, Bookie's Choice. "He won some minor races in his day," Peter said. "He's thirty years old. He's a nice little guy."

But, he is not even close to being the oldest. "We have some horses who are very old," Mary said. "We have a horse who is thirty-nine."

The staff at the home that takes care of the horses consists of Peter, Mary, one full-time employee, one part-time employee, and a handful of volunteers. A veterinarian, who is on the home's board of directors, and a farrier provide their services at a discounted price.

In the spring of 2004, the Gregorys were contacted to adopt a horse found in the Everglades. "Apparently, people dump their animals, when they don't want them, in the Everglades," Mary said. "Some have been tied to a tree. Either the alligators get them or wild dogs. Someone sent an article to us. It's only coming to light now. They're cleaning up the Everglades and they're finding skeletons of cows and horses."

The home is often a haven for such castoffs. "There aren't many people who do what we do," she said. "Other rescue places get the horse well

Peter and Mary Gregory at their Retirement Home for Horses in Florida with Willa, rescued from the Everglades.

again and adopt them out. When the horse comes through our gate, we promise it will never leave the property and it never does."

Recently, the city of Miami and the counties of Dade, Broward, and Palm Beach have been supplying a steady number of workhorses to be retired. So has the American Society for the Prevention of Cruelty to Animals. The Peter and Mary Gregory receive the horses with open arms. "There are no requirements from us, only that we agree not to sue the city for any reason," Peter said. "The city of Miami usually sends ten bags of feed."

In 2002, the Gregorys were honored with plaques in a special ceremony at the City of Miami Police Department headquarters. "There are no words to express how grateful this department is that such a place like Mill Creek Farm exists," said Terry Pagan, the barn supervisor of the mounted police unit.

A perpetual conservation easement on the land will ensure the home continues as is, even after the Gregorys are gone. Said Peter:

They make sure that twenty years after we're dead, the land remains the way it is and is used for horses. The land can't be broken up or sold. We can die with a clear hope this thing will continue.

I always look at it as we're lucky enough to be able to do it. We come home exhausted every day, but it's a feeling of exhilaration. There's nothing more beautiful than going out every morning and feeding the horses. It's like being in heaven.

He and his wife have saved a lot of horses from hell.

APPENDIX

SAVING HORSES

Programs in the United States and Canada Saving, Retiring, and/or Placing Horses, Donkeys, and Burros

American Horse Defense Fund
Trina Bellak
866-893-3456
www.ahdf.org
AHDForg@aol.com

Blue Horse Charities
Leslie McCammish
859-255-1555
www.bluehorsecharities.org
imccammish@fasigtipton.com

CANTER
Jo Anne Normile
734-455-0639
www.canterusa.org

Equine Advocates
Susan Wagner
845-278-3095
www.equineprotection network.com

Equine Protection Network
Christine Barry
www.equineprotection network.com

HoofPAC.com
Cathleen Doyle
www.hoofpac.com

Humane Society of U.S.
818-501-2275
www.hsus.org

National Horse Protection Coalition
202-33-4936
www.horse-protection.org

Racehorse Adoption Referral Program
877-576-NTRA
www.NTRA.com

ReRun
Shon Wylie
859-289-7786
www.rerun.org

Thoroughbred Charities of America
Terry Conway
302-427-0905
www.thoroughbred charities.org

United States

ALABAMA

Alabama Equine Rescue
205-680-1862
www.aerescue.netfirms.com
AERescue2000@ yahoo. com

Foal Train
Liz Creamer
251-545-7980
www.foaltrain.com
tugonurhrt@aol.com

Magic King & Friends Rescue
www.expage.com/ rescuefarm

ALASKA

Alaska Equine Rescue
Sally Clampitt
888-588-4677 or 888-LUV-HORSE
www.alaskaequineres cue.com
aer@alaskaequinerescue.com

ARIZONA

Hacienda de los Milagros
Wynne Zaugg
520-636-5348
www.haciendadelosmila gros.org

Hooved Animal Humane Society
www.hahsofaz.com
hahsofazcv@yahoo.com

Maricopa County Horse Rescue
623-434-5602
www.werescuehorses.com

Painted Promise Ranch
Randy Morgan
623-388-2068
www.paintedpromise.com
rescue@paintedpromise.com

Rescue A Horse.com
Holly Marino
602-689-8825
www.rescueahorse.com
holly@rescueahorse.com

TheHorseRescue@aol.com
Mea Equine Sanctuary
www.dhhrescue.com

Wildhorse Ranch
408-926-1795
www.mudpony.com
mudpony@aol.com

X-S Ranch Livestock Rescue & Sanctuary
www.xs-ranch.20m.com

ARKANSAS

Ozland Horse Rescue
http://myozland.tripod. com/ozlandhorserescue/

ReRun
Jennifer Keller
501-997-8581
www.rerun.org
jkeller@rerun.org

CALIFORNIA

Alternate Careers for Thoroughbreds

Noelle Poulsen
www.geocities.com/act
_California

California Equine Retirement Fund
Grace Belcuore
909-926-4190
www.cerfhorese.orgcerf
@pe.net

Equus Sanctuary
530-931-0108
www.equus.org

Friends of Arabian Horse Rescue
760-240-8928
ArabianRescue@mail.
com

Glen Ellen Vocational Academy (GEVA)
Pam Berg
707-527-8092
www.glenellenfarms.com
/gevagef@vom.com

Golden Carrot
Casey O'Conner
909-763-0800
www.home.earthlink.net/
~cocarrot
cocarrot@earthlink.net

Horses' Honer
Nina Clark
530-823-8927
www.geocities.com/
horses_honor

Jack Auchterlonie Memorial Equine Sanctuary (James)
Fran Becker
760-362-1357
www.jamesrescue.com
james29palms@aol.com

Lasso Horse Rescue
Larry Simms
970-264-0095
www.lassohorserescue.org
lassohorserescue@yahoo.com

Lifesavers (Mustangs)
661-727-0049
www.wildhorserescue.org

Meadowbrook Animal Sanctuary
Renee Duncan
909-245-8661
www.mashresuce.orgen
challa@aol.com

Mustang Rescue Ranch
ffillycga@aol.com

Painted Spirit Ranch
Kim Reynolds
909-677-5464
www.paintedspiritranch.
com

Phoenix Equine Foundation
Teri Rohm
707-833-4774
Jeanie@ap.net

Redwings Horse Sanctuary
831-286-0195
www.redwings.org
info@redwings.org

Return to Freedom
Neda De Mayo
805-737-9246
www.returntofreedom.org
info@returntofreedom.org

Southern California Equine Foundation
Leigh Gray
909-590-0988
www.scef-inc.com

Tranquility Farm
Priscilla Clark
661-823-0307
www.tranquilityfarmtbs.org

True Innocents Equine Rescue
909-360-1464
www.tierrescue.org
info@tierrescue.org

United Pegasus Foundation
Helen Meredith
626-279-1306
www.unitedpegasus.com
unitedpegasus@yahoo.com

Wild Burro Rescue
Diana Chontos
760-764-2136
www.wildburrorescue.org

Wild Horse Sanctuary
Dianne Nelson
530-335-2241
www.wildhorsesanctuary.com
whs3@frontiernet.net

COLORADO

Colorado Horse Rescue
Nan Millett
720-494-1414
www.chr.org

Friends of Horses Rescue & Adoption
Bill Stiffler
877-229-7483
www.getahorse.org

Front Range Equine Rescue
Hillary Wood
719-495-6389
www.frontrangeequine
rescue.org

Horse Protection League
Bini Abbott
303-424-0037

Lasso Horse Rescue
Larry Simms
970-264-0095
www.lassohorserescue.org

Longhopes Donkey Shelter
Kathy Dean
303-644-5930
www.longhopes.org

Phoenix Rising Ranch
719-351-9862
www.phoenixrescue.org

Rocky Mountain Foal Rescue
Pam Pietsch
719-683-5880
www.rockmountainfoal
rescue.org
info@7starsanctuary.org

CONNECTICUT

Citizens for Horse Protection
capinc@usa.net

Greener Pastures Rescue
www.greenerpastures
rescue.org
4asteed@steedread.com

HORSE of Connecticut
Patty Wahlers
860-868-1960
horseofct@yahoo.com

New Haven Animal Rescue Association
care4animals@mail.com

Thoroughbred Retirement Foundation
Melissa Klick
800-728-1660
www.trfinc.org

DELAWARE

Lovers United
Lorraine Truitt
410-749-3599
www.horseloversunite.com

Summer Wind Stables
Elena DiSilvestro
877-SMRWIND
www.summer windstables.com

Tri State Equine Rescue and Adoption
302-492-0492
www.tristateequine.org

FLORIDA

Adopt-A-Horse
Judy Detore
863-382-4483
www.adoptahorse.com

Animal Rescue Kingdom
Diane Zandman
352-291-1678

Back in the Saddle Horse Adoption (BITS)
Pam Rayl
904-284-5852
www.bitshorseadopt.org
info@bitshorseadopt.org

Equine Rescue & Adoption Foundation
561-627-1198
www.equinerescuefl.org
info@equinerescuefl.org

Friends of the EIA Horse
Debbie Barwick
954-492-0168
www.eiahorses.org

Heavenly Meadows Horse Rescue
Vicky Hammond
850-773-9991
www.heavenlymeadows.org
info@heavenlymeadows.org

HOPE
Tara McDonald
813-417-6386
hope4tbreds@aol.com

Hope for Horses
772-785-5309
www.hopeforhorses.net
info@hopeforhorses.net
hope4horses@hotmail.com

Horse Protection Association
Morgan Silver
352–466-4366
www.hpaf.org

Retirement Home for Horses
Peter Gregory
386-462-1001
www.millcreekfarm.org
rhh@millcreekfarm.org

South Florida SPCA
Roy Pressman
305-825-8826
www.spca-sofla.org

Standardbred Pleasure Horse Organization of Florida (SPHO of Florida)
Debra Swegar
352-796-4842
ww.sphofl.com
dswegar@juno.com

Thoroughbred Retirement Foundation
Melissa Klick
800-728-1660
www.trfinc.org

Whispering Winds Ranch
352-472-3925
www.whisperingwinds ranch.homestead.com
taini2727@aol.com

GEORGIA

Begin Again Farms Equine Shelter
Rhonda Jackson
www.beginagain farms.com
Beginagainfarms@aol.com

Exceller Fund to Rescue Horses
Karla Phaneuf
770-998-7368
www.excellerfund.org

Georgia Equine Rescue League
Marty Paulk
770-464-0138
www.gerlltd.org
info@gerlltd.org

Horse Rescue, Relief & Retirement
Cheryl Flanagan
70-886-5419
www.SaveTheHorses.org

IDAHO

Orphan Acres
208-882-9293
orphan_acres@hot mail.com

ILLINOIS

CANTER
Denise Fillo
630-850-5548
www.canterusa.org

Chance's Mini Horse Rescue
Kay Baxter
217-784-1307
www.chancesminihorse rescue.org
kay@chancesminihorse rescue.org

Crosswinds Equine Rescue
217-832-2010
www.crosswindseqresq.org

Hooved Animal Humane Society
Lydia Miller
815-337-5563
info@hahs.org

ReRun
Gingy Cody
847-428-9371
www.rerun.org

INDIANA

Indiana Horse Rescue
765-659-5209
www.esfrescue.com

Moon Winds Rescue
765-676-5798
www.moonwinds
rescue.com

IOWA

Lazy R Equine Rescue
Karen Ralston
563-652-4593
www.angelfire.com/ia2/
lazyriowa/main.html
ranny2000@earthlink.net

Rollin' on Racehorse Rescue
Don Buboltz
www.racehorserescue.com

KENTUCKY

Casey Creek Horse Rescue & Adoption Inc.
Kenneth Holland
270-789-4198
www.homestead.com/
cchra/cchra.html
desperado_55@yahoo.com

Dream Catcher Stables
606-376-2533
www.dreamcather
stables.net

Equine Transitional Training
Keri Basham
859-846-5655
ettahorse@hotmail.com

Old Friends
Michael Blowen
859-846-9995
www.oldfriendsequine.org

ReRun
Shon Wylie
859-289-7786
www.rerun.org
rerunhorse@kih.net

Witch Trot Farms
502-839-4425
remarking@aol.com

LOUISIANA

Hopeful Haven Equine Rescuer
Debra Barlow
318-92-4272
www.hopefulhaven.com

MAINE

Standardbred Pleasure Horse Organizaion
www.sphomaine.net

MARYLAND

Days End Farm Horse Rescue
Kathy Schwartz
301-854-5037
www.defhr.org

Equine Rescue & Rehabilitation
Deborah Fran
410-343-2142
www.horserescue.com

Fox Shadow Foundation
Jeannie Meade
410-673-2634
foxshadow@dmv.com

Gambler's Pastures Equine Sanctuary
Vera Karelian
410-827-4766

Greener Pastures Equine Sanctuary
Wendy Moon
410-885-2198
www.thoroughbred
charities.org

Horse Lovers United
Lorraine Truitt
410-749-3599
www.horselovers
united.com

HorseNet Horse Rescue
410-795-8989
www.horsenethorse
rescue.org

Lost & Found Horse Rescue
Kelly Young
717-428-9701
www.lfhr.org

MidAtlantic Horse Rescue
Beverly Strauss
302-376-7297
www.midatlantichorse
rescue.org

New Life Equine Rescue
310-305-0702
www.nler.org

Royal Equine Rescue & Sanctuary Inc.
Alyssa Taylor
443-417-0069
www.rersi.org
alyssa@rersi.org

The Keep at Andelain Farm Inc.
Wendy Moulton
301-271-4191
www.horsessavingkids.org
thekeep@andelainfarm.com

Thoroughbred Retirement Foundation
Melissa Klick
800-728-1660
www.trfinc.org

MASSACHUSETTS

CANTER
Ellen O'Brien
781-354-6291
www.canterusa.org/
newengland

Eye of the Storm Equine Rescue
Nina Arbella
978-897-8866
www.equine-rescue.com

Suffolk Downs
Lori Lindermann
617-567-3900
www.suffolkdowns.com

MICHIGAN

Broken Pine Ranch Horse Rescue
brokenpineranch
@myway.com

CANTER
Jo Anne Normile
734-455-0639
www.canterusa.org

Great Lakes Hooved Animal Rescue
616-837-1144
www.hoovedrescue.org

Horses' Haven
517-548-4880

www.ismi.net/horses haven/index.htm

Hugs2Horses Inc.
517-223-3263
www.hugs2horses.com

Manitou Equine Rescue & Placement
Rita Hoag
Rubys@chartermi.net

Second Career Racehorses
Kris Kotlarz
616-913-2700
secondcareer racehorses.org

Second Chance Thoroughbred Adoption
Dale Berryhill
www.horsenetwork.com/ second chance/

MINNESOTA

A & L Equine Sanctuary
Laurie Parsons
218-397-2397
www.aandlequine sanctuary.com

Midwest Horse Adoption Program
mwhorseadoption @yahoo.com

Minnesota Hooved Animal Rescue and Rehabilitation Station
763-856-3119
www.mnhoovedanimal rescue.org

Misfit Acres
Bruce and Jody Thisted
507-278-4876
www.misfitacres.com

Save Our Souls Equine Rescue (SOSER)
28-637-2168
www.soser.usrescue@soser.us

MISSOURI

D-D Farm, Animal Sanctuary
Deb Tolentino
vhatfie@fableview.org

Fableview Equine Rescue
Valerie O'Brien
660-747-0531
www.fableview.org

Pientka Horse Rescue
Cheryl Pientka
816-690-7442

Rainbow Ridge Ranch Horse Sanctuary
www.rainbowridgeranch. org
info@rainbowridgeranch. org

The Original Funny Farm
www.geocities.com/ originalfunnyfarm

The Other Animals Rescue
Theanimalsrescue@aol.com

Thoroughbred Retirement Foundation
Melissa Klick
800-728-1660
www.trfinc.org

MONTANA

Angels Among Us Equine Rescue
Michelle Sudan
406-249-5016
pinball@centurytel.net

Montana Large Animal Sanctuary
406-741-3823
www.mtanimal sanctuary.com

NEVADA

Miracle Horse Rescue
775-751-1101
www.miraclehorse.com

Virginia Range Wildlife Protection Assoc.
775-881-2288
www.vrwpa.org

NEW HAMSHIRE

CANTER
Ellen O'Brien
781-354-6291
www.canterusa.org/ newengland

Live and Let Live Farm
Teresa Paradise
603-798-5615
www.liveandletlive farm.org

Thoroughbred Retirement Foundation
Melissa Klick
800-728-1660
www.trfinc.org

Turtle Rock Rescue
Ken Terry
603-585-9995
www.equineextra.com

NEW JERSEY

Enchanted Acres
732-730-1670

Lost & Found Horse Rescue
Kelly Young
717-428-9701
www.lfhr.org

Mylestone Equine Rescue
Susan Thompson
908-995-9300
www.mylestone.org

ReRun
Laurie Lane
732-521-4752
www.rerun.org

Standardbred Retirement Foundation
Barbara Budd
732-462-8773
www.adoptahorse.org

Thoroughbred Retirement Foundation
Diana Pikulski
800-728-1660
www.trfinc.org

NEW MEXICO

Horse Stories
Carolyn M. Bertin
505-281-9675
www.horsestories.org

Perfect Harmony Animal Rescue
505-824-2130
www.perfectharmony. bizland.com

The Horse Shelter
Jan Bandler
505-984-3235
www.thehorseshelter.org

NEW YORK

Balanced Innovative Teaching Strategies
Lynn Cross
518-794-8104
www.h-o-r-s-e.org
lynn@h-o-r-s-e.org

Blue Springs Farm Thoroughbred
Elena DiSilvestro
bluesprings@frinternet.net

Crane Mountain Valley Horse Rescue
Nancy Van Wie or Eddie Mrozik
518-623-3470
www.cmvhr.org

Eqine Advocates
Susan Wagner
518-245-1599
www.equineadvocates.com

Equine Rescue of Orange County
Lynda Broas
845-733-6085
www.equinerescue.org

Equine Rescue Resource
Colleen Segarra
845-744-1728
www.equineresecue resource.com

HiHopes for Horses
Donna O'Leary
www.hihopes.com

H.O.R.S.E. Rescue and Sanctuary
Chris Dodge
585-584-8210
www.hrsny.org

New York Horse Rescue
Mona Kanciper
631-874-9420
mona@nyhr.org

ReRun
Sue Swart
315-598-2898
www.rerun.org

Tender Mercy Equine
716-471-4796
www.tendermercy rescue.com

Thoroughbred Retirement Foundation
Melissa Klick
800-728-1660
www.trfinc.org

Western New York Equine Sanctuary
CarolAnn Piazza
716-438-0182
cpiazza912@aol.com

NORTH CAROLINA

Carolina Equine Society
Stephen Hill
336-768-1514
www.carolinaequine society.org

Horse Protection Society of North Carolina
Joan Benson
704-855-2978
www.horseprotection.org

United States Equine Rescue League
Jacalyn Ackerman
336-720-9757
www.userl.org

NORTH DAKOTA

Sleipnir Horse Sanctuary
Scott and Ame Smith
701-782-4217
http://sleipnirsanctuary.tripod.com

OHIO

CANTER
Nancy Scott
330-697-4460
www.canterusa.org/ohio

Happy Trails
Annette Fisher
330-296-5914
www.happytrailsfarm.org

Humane Retired Standardbreds
Suzanne Stevens
419-768-2255

JT's Horse Project
419-385-5315
jtshorseproject@msn.com

Last Chance Corral
Victoria Goss
740-594-4336
www.lastchancecorral.org

Lost Acres Horse Rescue & Rehabilitation
Sissy Burggraf
740-779-6761
www.geocities.com/sblahrr

New Vocations Racehorse Adoption
Dot Morgan
937-947-4020
www.horseadoption.com

Second Chance Horse Rescue
Daniel's Thoroughbred Farm
Alice Crow
513-200-0808
www.secondchancehorse rescue.com
info@secondchancehorse rescue.com

Thoroughbred Retirement Foundation
Melissa Klick
800-728-1660
www.trfinc.org

OKLAHOMA

Thoroughbred Retirement Foundation
Melissa Klick
800-728-1660
www.trfinc.org

OREGON

Emerald Valley Equine Assistance
Sandy Huey
541-935-3906
www.eveahr.com

Equine Angels
541-874-3517
equineangels@aol.com

PENNSYLVANIA

Angel Acres Horse Haven Rescue
Jo Deibel

717-225-3212
www.angelacreshorsehaven
rescue.com

Another Chance for Horses
Kathy George
434-384-9465
www.anotherchance
forhorses.com

Bran Manor Equine Rescue
Tru Marabella
610-587-1098
www.angelfire.com/ri2/
branmanorrescue

Bright Futures Farm
Beverlee Dee
814-827-8270
www.giddiupgo.com/hand
somehorses/home.shtml

Down Patrick Farm
Lisa Popp
717-567-7164
lpopp@pa.net

High Hope
Joa Haas
610-273-752
lizajoa@aol.com

Hooved Animal Welfare Council
Elena DiSilvestro
814-899-0960
majek25@hotmail.com

Last Chance Ranch
Lori Benetz
215-538-2077
www.lastchanceranch.org

Lost & Found Horses Rescue
Kelly Young
717-428-9701
www.lfhr.org

Paradise Equine Adoption & Youth
724-775-4470
www.paradiseequine
adoption.org

Rainbow Acres Horse Rescue
hapbhors@juno.com

Ryerss Farm for Aged Equines
Joseph Donahue
610-469-0533
www.ryerss.com

Whispering Acres
Dan Beachel
570-966-9374
www.geocities.com/
whisperingacres

Wind Ridge Farm Equine Sanctuary
Gary Barnes
717-432-2959

SOUTH CAROLINA

PEER
Bernie Peeples
843-871-1909
www.peerhorserescue.org

South Carolina Awareness Rescue
Karen Metze
888-866-8744
schorserescue.org

SOUTH DAKOTA

Black Hills Wildhorse Sanctuary
Dayton Hyde
www.wildmustangs.com

TENNESSEE

Angel Rescue and Transport
Lena M. Frensley
615-740-0964
www.angelrescue.com
LenaFrensley@angelrescue.com

Horse Haven of Tenessee
Nina Marngetson
865-609-4030
www.horsehavenoftn.com

TEXAS

Animals First
Holly Christian
936-228-0434
www.animals1st.com
animals1strescue@
hotmail.com

Brighter Days Horse Refuge
Jeanie Weatherholz
830-510-6607
www.brighterdays
horserefuge.org

Equiserve Equine Rescue
Janice McCoy
817-220-0775
www.equiserveequine
rescue.org
jlmccoy@equiserveequine
rescue.org

Exceller Fund to Rescue Horses
Barbara Kraus
972-874-7486
www.excellerfund.org

Habitat for Horses
409-935-0277
www.hopeforhorses.org

Hope for Horses
972-734-6218
www.hopeforhorses.org

Humane Help Animal Rescue
432-229-4295
www.freewebs.com/hhar/

Last Refuge
Carol Chapman
972-962-7706
www.geocities.com/
~lastrefuge/

Lone Star Equine Rescue
Jenny Williams
979-776-9396
www.lser.org

Sam Houston Race Park Adopt-a-Horse
281-807-8700
www.shrp.com

Shady Oaks Ranch
LeAnn Souser
817-648-0122
www.shadyoaksranch.com
leann@shadyoaksranch.com

Southwestern Equine Rescue Foundation
Christy Clements
469-384-0230
cclements@earthlink.net

United States Equine Sanctuary & Rescue
Daniele Pruitt
877-720-1685
www.usesr.org
dpruitt@usesr

VERMONT

Humane Organizaion for Un-Raceables
Terry Holden
802-644-2759
HORSEVT@sover.net

Spring Hill Horse Rescue
Gina Brown
802-247-2857
Springhillrescue@aol.com

Thoroughbred Retirement Foundation
Melissa Klick
800-728-1660
www.trfinc.org

VIRGINIA

Dream Catcher Farm Horse Sanctuary
www.horsesanctuary.com

Equine Rescue League
Pat Rogers
www.equinerescueleague.org

Horse Lovers United
Lorraine Truitt
410-749-3599
www.horselovers united.com

Lost & Found Horse Rescue
Kelly Young
717-428-9701
www.lfhr.org

ReRun
Paul Linn
434-369-5433
www.rerun.org

Roanoke Valley Horse Rescue
Pat Gillar
540-797-1999
www.rvhr.com

The Laughing Horse Sanctuary
Tom and Julia Durfee
434-927-5298
www.laughinghorse.org

Thoroughbred Retirement Foundation
Melissa Klick
800-728-1660
www.trfinc.org

WASHINGTON

Broken Oaks Equine Retirement Center
Jean and Gary Pratt
509-767-1461
www.brokenoaks.org
gpratt@gorge.net

Hooved Animal Rescue of Thurston County
Joyce Bell
www.har-otc.com

People Helping Horses
360-920-6444
www.peoplehelping horses.com

ReRun
Jody Peetz
425-313-5982
www.rerun.org

Sleepy Hollow Horse Rescue
Sarah Schak
509-488-7669
www.sleepy-hollow.org

Whidbey Island Rescue of Equines
Becky Montoya
360-675-9252
www.petfinder.org/shelters/WA151.html

WEST VIRGINIA

CANTER
Allison Conrad
301-728-6062
www.canterusa.org

C & M Equine Rescue
Michelle Eddy
www.cmranch.org

Lost & Found Horse Rescue
Kelly Young
717-428-9701
www.lfhr.org

Second Wind Adoption Program
304-873-3121
www.crossedsabers.com
secondwindadopt@aol.com

WISCONSIN

American Standardbred Adoption
Susan Wellman
608-637-8045
www.4thehorses.com

Central States Horse Rescue
Laurie Parsons
www.cshr.50megs.com

Midwest Horse Welfare Foundation
Scott Bayerl
715-387-0555
www.equineadoption.com
yukon@tznet.com

Prairie-Woods Horse Rescue
715-246-5323
info@prairie-woods.org

Thoroughbred Retirement Foundation
Melissa Klick
800-728-1660
www.trfinc.org

CANADA

Foalquest
Wendy Ellis
www.pmufoalquest.com

Heaven Can Wait
Claire Malcolm
705-359-3766
www.heavencan
waitequinerescue.org

LongRun Thoroughbred Retirement
Vicki Pappas
416-675-3993
www.longrun
retirement.com

NetPets
Karen Briggs
905-352-2662
www.netpets.com

New Stride Thoroughbred Retirement
Cathy Sheppard
604-255-7652

Ontario Standardbred Adoption
Kelly McLaughlin
905-812-7408
www.digitalwave.ca/osas

SOS PMU Foal Rescue
Christina Rae
http://members.shaw.ca/
sosrescue

INDEX

ABOUT THE AUTHOR

Bill Heller is a frequent contributor to *Thoroughbred Times* and *Mid-Atlantic Thoroughbred.* Heller also handicaps thoroughbred racing for the *Daily Gazette* in Schenectady, N.Y. He has received nine national awards, including the 1997 Eclipse Award for Outstanding Magazine Writing on Thoroughbred Racing, the 1999 Bill Leggett Breeders' Cup Magazine Writing Award, and the John Hervey Award for outstanding magazine writing about harness racing (three times). Heller is a member of the National Turf Writers Association, New York Turf Writers Association, and the United States Harness Writers Association. Heller has written seventeen other books, including several racing titles, such as *Go for the Green*; *Graveyard of Champions*; *Run, Baby, Run*; and *Saratoga Tales.*

A native of Liberty, N.Y., Heller attended Albany State University and graduated with honors with a major in political science and minor in journalism. He lives in Albany, N.Y., with his wife, Anna, their son, Benjamin, and their dog, Belle Mont.

Other Books by Bill Heller

Obsession: Bill Musselman's Relentless Quest; Overlay, Overlay; The Will To Win: Ron Turcotte's Ride to Glory; Playing Tall—The Ten Shortest Players in NBA History; Harness Overlays; Travelin' Sam—America's Sports Ambassador; Exotic Overlays; Billy Haughton—The Master; Turf Overlays; Thoroughbred Legends: Go for Wand; Thoroughbred Legends: Forego; Thoroughbred Legends: Personal Ensign; Graveyard of Champions; Run, Baby, Run; A Good Day Has No Rain; Go For the Green; Saratoga Tales